T0256645

Life Out of Sequence

Life Out of Sequence:
A Data-Driven History of Bioinformatics

Hallam Stevens

The University of Chicago Press :: Chicago and London

Hallam Stevens is assistant professor at Nanyang Technological
University in Singapore.

The University of Chicago Press, Chicago 60637
The University of Chicago Press, Ltd., London
© 2013 by The University of Chicago
All rights reserved. Published 2013.
Printed in the United States of America

22 21 20 19 18 17 16 15 14 13   1 2 3 4 5

ISBN-13: 978-0-226-08017-8 (cloth)
ISBN-13: 978-0-226-08020-8 (paper)
ISBN-13: 978-0-226-08034-5 (e-book)
DOI: 10.7208/chicago/9780226080345.001.0001

Library of Congress Cataloging-in-Publication Data

Stevens, Hallam.
    Life out of sequence : a data-driven history of bioinformatics /
Hallam Stevens.
        pages. cm.
    Includes bibliographical references and index.
    ISBN 978-0-226-08017-8 (cloth : alk. paper) — ISBN 978-0-226-
08020-8 (pbk. : alk. paper) — ISBN 978-0-226-08034-5 (e-book)
    1. Bioinformatics—History. I. Title.
QH324.2.S726 2013
572'.330285—dc23

                                                    2013009937

♾ This paper meets the requirements of ANSI/NISO Z39.48-1992
(Permanence of Paper).

For my parents

If information is pattern, then noninformation should be the absence of pattern, that is, randomness. This commonsense expectation ran into unexpected complications when certain developments within information theory implied that information could be equated with randomness as well as pattern. Identifying information with *both* pattern and randomness proved to be a powerful paradox, leading to the realization that in some instances, an infusion of noise into a system can cause it to reorganize at a higher level of complexity. Within such a system, pattern and randomness are bound together in a complex dialectic that makes them not so much opposites as complements or supplements to one another. Each helps to define the other; each contributes to the flow of information through the system.

N. KATHERINE HAYLES, *How We Became Posthuman: Virtual Bodies in Cybernetics, Literature, and Informatics*

I put my faith in randomness, not order.

J. CRAIG VENTER, *A Life Decoded: My Genome: My Life*

Contents

Introduction

> Will the domestication of high technology,
> which we have seen marching from triumph to
> triumph with the advent of personal computers
> and GPS receivers and digital cameras, soon be
> extended from physical technology to biotech-
> nology? I believe that the answer to this ques-
> tion is yes. Here I am bold enough to make a
> definite prediction. I predict that the domesti-
> cation of biotechnology will dominate our lives
> during the next fifty years at least as much as
> the domestication of computers has dominated
> our lives during the previous fifty years.[1]

Freeman Dyson wrote these words in 2007. Dyson was
one of the physicists who laid the groundwork for the
transformation of physics and physics-based technology in the second half of the twentieth century. He is not
alone in suggesting that biology will be to the twenty-first
century what physics was to the twentieth. Over the last
several decades, biology has already been transformed.
The growing importance of bioinformatics—computers,
data, and data management—signals that fundamental
changes have taken place in how biologists work and
how they build knowledge.

But the great success and the ultimate "domestication" of physics in the second half of the twentieth century came at a significant price. The discipline was overhauled to an extent that would have made it hardly recognizable to practitioners of just a few decades earlier. Physics was transformed by the need for more money, more manpower, and more machines as well as by military involvement.[2] Physicists struggled with their changing identity as their work became increasingly oriented toward the management of large interdisciplinary teams working at centralized institutions. What it meant to *do* physics was fundamentally transformed over the middle decades of the century—practices, institutions, reward structures, career paths, education, and the core knowledge of the discipline were altered.

This transition did not occur without a struggle. Some physicists—and some non-physicists—worried intensely about what the advent of Big Science would mean not only for their own professional futures, but also for the future of scientific knowledge. Although it was Donald Glaser's invention of the bubble chamber that provided the model for the first large-scale postwar physics laboratories, he was so disturbed by the new culture of physics that he quit the discipline entirely. Glaser longed for the days of the individual scientist working diligently at his lab bench and disliked the corporate, collective world that physics had become.[3]

In his valedictory address, President Dwight Eisenhower famously worried about the corrupting effects that large amounts of money would have on the purity of scientifically produced knowledge; the "military-industrial complex," he fretted, posed an immediate threat to free ideas and intellectual curiosity:

> Today, the solitary inventor, tinkering in his shop, has been over-shadowed by task forces of scientists in laboratories and testing fields. In the same fashion, the free university, historically the fountainhead of free ideas and scientific discovery, has experienced a revolution in the conduct of research. Partly because of the huge costs involved, a government contract becomes virtually a substitute for intellectual curiosity. For every old blackboard there are now hundreds of new electronic computers.[4]

Alvin Weinberg, director of the sprawling Oak Ridge National Laboratory, which had helped to build the first atomic bombs, was also concerned about the effects of Big Science on both scientists and society. These effects were felt most strongly by the individual investigator:

The research professor in a simpler day concerned himself with the substance of his science, both in research and teaching. Now, through no fault of his own, he must concern himself with many other matters. To do his research, he has to manage, even in Little Science, fairly large sums of government money. He must write justifications for his grants; he must serve on committees that recommend who should receive support, who should not; he must travel to Washington either to advise a government agency or to cajole a reluctant contract administrator. In short, the research professor must be an operator as well as a scientist.[5]

Such changes, Weinberg argued, led to reduced flexibility in scientific practice, compromised its "intellectual discipline," created increasing fragmentation between the specialized branches of science, and forced difficult decisions about what kinds of science to support. Weinberg particularly criticized high-energy physics—the apotheosis of Big Science—for its failure to address problems of human welfare and technology and for its remoteness even from other branches of physics.[6] Those who worked in less well funded subdisciplines, such as solid-state physics, wondered whether high-energy physics really deserved so much cash. In 1972, Philip W. Anderson argued that the "arrogance" of particle physics was causing important domains of investigation to be under-studied and underfunded; solid-state physics presented an alternative to the reductionist approach of Big Science.[7] Such arguments lay at the surface of a deeper pool of discomfort about what money, machines, and the military had done to physics. Big Science forced reflection on the most appropriate ways to make physical knowledge.[8]

Biology, as it too becomes Big Science, is currently experiencing many of the same changes that revolutionized physics in the mid-twentieth century. Significantly, many of these changes are being driven by the computerization of biological work and knowledge. Eisenhower's image of blackboards being transformed into electronic computers reminds us that computers not only played a crucial role in transforming physics, but also became an important symbol of the new styles of working and knowing. Information technologies—computers, networks, robotics—are playing an even more central role in transforming the practices of biology in the early twenty-first century.

What did a biology laboratory look like before computers, Big Science, and genomics? The ethnographers Bruno Latour and Steve Woolgar gave a firsthand account of a neuroendocrinology lab from the 1970s.

They described workers preparing assays, attending to instruments, and recording data, as well as the flows of paper (mail, journal offprints, and typed manuscripts), animals, and chemicals through the lab.[9] No doubt there are many laboratories that still more or less conform to this description: work is organized around benches, which are filled with reagents and samples, technicians and scientists communicate and collaborate with one other primarily in face-to-face interactions around blackboards or whiteboards, and the most important things that come in and out of the laboratory are physical materials and printed papers.

But by the mid-1980s, changes were afoot. In 1991, Walter Gilbert, one of the inventors of DNA sequencing, proclaimed the coming of a "paradigm shift" in molecular biology: "Molecular biology is dead— long live molecular biology!"[10] The old "experimental" paradigm was to "identify a gene by some direct experimental procedure—determined by some property of its product or otherwise related to its phenotype, to clone it, to sequence it, to make its product and to continue to work experimentally so as to seek and understanding of its function." Gilbert worried that this "list of techniques" was making molecular biology more technology than science—it would become "pocket molecular biology" that could be performed by reading a recipe book or purchasing a kit.[11]

His answer to this was a new paradigm in which

> all the "genes" will be known (in the sense of being resident in databases available electronically), and . . . the starting point of a biological investigation will be theoretical. An individual scientist will begin with a theoretical conjecture, only then turning to experiment to follow or test that hypothesis.[12]

Gilbert emphasized the need for biologists to become computer literate, to "hook our individual computers into the worldwide network that gives us access to daily changes in the database," and thus to take advantage of the overwhelming amounts of data to "change their approach to the problem of understanding life."[13] Just as the Human Genome Project was gearing up, Gilbert perceived a change toward a highly data-driven and even "theoretical" approach to biological problems.

But the kinds of problems and challenges Gilbert identified were just the tip of the iceberg. Just like in physics, the transition to computerized Big Biology required new institutions, new forms of organization, new ways of distributing resources, new forms of practice, and new kinds of knowledge.

When I ventured into a laboratory in the Department of Biology at MIT in the summer of 2007, the scene looked remarkably different from the lab of the 1970s. The days began slowly as graduate students and postdocs wandered into the laboratory, exchanging greetings but nothing more. The coffee machine was switched on. In half of the lab, benches had been replaced by long rows of desks, each holding three or four computers. Each member of the lab gravitated to his or her own workstation and resumed his or her own project. Occasionally, workers would leave their desks to refill their cups of coffee. Technicians and secretaries were nowhere to be seen. Few other physical objects ever moved into or out of the laboratory. Each desk sat adjacent to a small whiteboard that individuals used to make notes or write reminders to themselves. Although the lab housed several printers, they were only rarely used to print papers and diagrams, almost never data.

Here, most of the flow into and out of the laboratory was virtual; data, articles, results, diagrams, and ideas moved through the space invisibly. The most obvious architectural division was no longer between the bench space and the office space, but between the "wet lab" (full of lab benches) and the "dry lab" (full of computers).

These differences in appearance and work demonstrate the fundamental changes that have taken place in biology in the last thirty years. Gilbert's paradigm shift began to change the meaning of the very objects of biology itself. That is, computers have altered our understanding of "life." In the first place, this change involved the "virtualization" of biological work and biological objects: organisms and genes become codes made up of zeros and ones. But more importantly, information technologies require particular structures and representations of biological objects. These structures and representations have increasingly come to stand in for the objects themselves in biological work. Databases and algorithms determine what sorts of objects exist and the relationships between them. Compared with the 1960s and 1970s, life *looks different* to biologists in the early twenty-first century.[14]

The wet labs and wet work of biology have not disappeared, but they are increasingly dependent on hardware and software in intricate ways. "Seeing" or analyzing a genome, to take one important example, requires automated sequencers, databases, and visualization software. The history recounted in this book is just not a story about how computers or robots have been substituted for human workers, or how information and data have replaced cells and test tubes in the laboratory. These things have occurred, but the changes in biology are far deeper than this. Nor is it just a story about how computers have speeded up or scaled up biology. Computers are implicated in more fundamental changes:

changes in what biologists do, in how they work, in what they value, in what experimentation means, in what sort of objects biologists deal with, and in the kind of knowledge biology produces. "Bioinformatics" is used here as a label to describe this increasing entanglement of biology with computers. By interrogating bioinformatic knowledge "in the making," we learn how biological knowledge is made and used through computers. This story is not about the smoothness of digital flows, but about the rigidity of computers, networks, software, and databases.

Putting it another way, the chapters of this book all attempt to answer the question, "What is bioinformatics?" from slightly different perspectives. Bioinformatics does not have a straightforward definition; it overlaps in complicated ways with other terms such as "computational biology" and "systems biology." It is a site of contestation about what biology will look like in the future: What forms of practice will it involve? How will knowledge be certified and authorized? What kinds of knowledge will count as biological? And in particular, what roles will computers and the digital play in biological knowledge? Bioinformatics entails competing answers to these questions. In this sense, bioinformatics is incomplete: it is a set of attempts to work out how to use computers to do and make biology. This is a pressing issue precisely because computers bring with them commitments to practices and values of Big Science: money, large institutions, interdisciplinarity. Bioinformatics is not just about whether or not to use computers to do biology, but also about how to organize biological practice and biological knowledge more generally.

Data Biology and Economies of Production

At the center of all these changes are data. Data constitute the (virtual) stuff created by experiments, reduced in analysis, moved and arranged inside the machine, and exchanged and communicated through networks. Data may be numbers, pieces of text, letters of a DNA sequence, or parts of an image. When biologists are using computers, they are dealing with data in some way. These data are not knowledge. A biologist can submit data to a database, but he or she cannot submit them to a journal for publication.[15] Their value lies in the fact that if they are appropriately treated and manipulated they can be used to create knowledge, but they are something other than knowledge.

Paying attention to data—rather than information—avoids suggesting that there is something inside the organismic body (code, information) that is somehow passed into the computer. Information is treated as an immaterial message encoded in the chemistry of DNA that is then extracted into computer codes. Data, on the other hand, do not flow freely out from or across biological bodies—they are tightly coupled to the hardware and software within which they exist. Data belong only to computers; they are part of a set of practices that make sense only with and through computers. Following the data helps us cut through the information metaphors and get closer to understanding the perplexity of things going on inside the machine.

Data, and especially "big data," have received much recent attention from scientists. Biologists, along with astronomers, physicists, and Google's engineers, have worried a great deal about how to extract knowledge from data. These concerns are centered not only on the need for new methods and techniques for dealing with large amounts of data, but also on the need for new institutions, new standards, new modes of work, new funding models, and new kinds of training in order to generate and support innovative methods of dealing with data. "Funding agencies have been slow to support data infrastructure," a *Nature* editorial reported in 2008, and "researchers need to be obliged to document and manage their data with as much professionalism as they devote to their experiments."[16] Finding ways to deal with data required new "incentives" for researchers.

Data are recognized as at once a technical and a social problem for science. As such, they provide a ready-made lens to see how both scientific practice and scientific knowledge are being transformed through their production and treatment. Computers are—first and foremost—machines for the storage, processing, and analysis of data. "Following the data," then, provides a way of getting inside the machines, seeing how they are connected, seeing how things move around inside them, and understanding what constraints they impose. This means not only seeing how data are produced and used, but also what structures they inhabit and what kinds of practices and skills are required to maintain them.

Social scientists have, for some time now, been paying attention to material culture. Historians of science in particular have shown convincingly how tools, objects, media, and so on have influenced the kind and character of scientific knowledge.[17] The premise on which this investigation is based is that to understand bioinformatics, we need to supplement our attention to materials (and material culture) with attention to data (and data culture). Attention to material culture has shown us that power and knowledge are embedded not just in people and documents, but also in objects. Data draw objects into new relationships and new shapes. And these data structures, movements, and patterns contain knowledge and exert force on behavior. The ways in which data are generated, used, and stored constitute cultures that might be described and analyzed.

To understand computerized science and the virtual worlds that it creates, we must look to the practical imperatives of making, managing, manipulating, storing, and sharing data. Michael Fortun expresses the importance of data for contemporary science through his notion of "care of the data":

> It is through the care of the data that many genomicists work out and realize what they consider to be good science, and "the good scientist": open to the data avalanche, welcoming its surprises and eagerly experimenting with them, but simultaneously cultivating a respect for the data flood's openness to multiple interpretations, a keen awareness of its uncertainties, fragilities, and insufficiencies, a heightened sensitivity to its confounding excesses and its limitations.[18]

What is at issue, however, is not only that biologists (and other scientists) must be increasingly attentive to the subtleties of data management, but also that data increasingly exert their own constraints on scientific knowledge making. Data constrain what can and cannot be done in biology. This book follows the data to find out where biology is going.

The importance and ubiquity of data in biology are already obvious. As Fortun points out, genomics in particular is concerned with amassing more data, analyzing them in high throughput, and storing them in ever-larger databases. It is a science already obsessed with data. The aim here is not to be swept up in this data flood, but to understand how data mediate between the real and the virtual and to analyze their consequences for biological knowledge. It is not that life "becomes" data, but rather that data bring the material and the virtual into new relationships. Following the data provides a way of locating, describing, and analyzing those relationships; it allows us to acknowledge that materiality is never fully erased, but rather, the material of the organism and its elements is replaced with other sorts of material: computer screens, electrons, flash memory, and so on.

Finally, paying attention to data can help us understand the economies of data production and exchange in biology. Just as Robert Kohler found that fruit flies generated a particular moral economy for *Droso-*

phila genetics, data biology produces particular economies of work and exchange.[19] Following the data is not only about studying things inside computers, but also about showing how the flow and organization of data—between people, through laboratories—affect the organization and practice of biological work.

Life Out of Sequence

Describing the ongoing changes in biology demands a mixed set of sources and methods. In some cases, a properly historical approach is appropriate; in others, a more anthropological account is required. Scientific publications, unpublished archival sources, interviews with biologists, and observations drawn from immersive fieldwork in laboratories all find a place.[20] The histories of computers, biological databases, and algorithms are all a part of this story, but it is not a straightforward history of bioinformatics. There are detailed accounts of laboratory practice, but it is not just an ethnography. Rather, a combination of historical and ethnographic methods is used to answer historical questions about how and why biology has changed due to computing.

Six chapters chart the different paths of data into and through biology. The first two chapters examine the introduction of computers into biology as data processing machines. Computers were developed as machines for data analysis and management in physics. This heritage is crucial for understanding how and why computers came to play the roles they did in biology. Thirty years ago, most biologists would not have given much thought to computers. The first chapter explains how and why this began to change in the 1980s. By 1990, a new subdiscipline had emerged to deal with increasing amounts of sequence data. Chapter 2 characterizes this new field. Computers have reoriented biology toward large-scale questions and statistical methods. This change marks a break with older kinds of biological work that aimed at detailed characterization of a single and specific entity (for instance, one gene, or one signaling pathway).

The third and fourth chapters examine how data move around in physical and virtual spaces. These movements constitute new modes of biological work. Drawing primarily on fieldwork in biological labs, these chapters describe how the exchange and movement of data mean authorizing and valuing knowledge in new ways. Data-driven biology necessitates labs and work oriented toward productivity and efficiency.

The next two chapters examine data structures and how data become biological knowledge through databases and visual representations. Chapter 5 details the history of biological databases, including the work of Margaret Dayhoff and the development of GenBank. It shows how databases act to structure and order biological knowledge. Chapter 6 draws on published scientific literature, interviews, and fieldwork to piece together an account of the role of images as ways of making data into knowledge in computerized biology.

All this is by no means a chronological account of bioinformatics. It does not locate a definite origin, begin the narrative there, and follow the story through to the end.[21] Instead, it presents several parallel accounts of bioinformatics, each with its own origins and unique trajectories. Bioinformatics is a manifestly interdisciplinary set of practices—it sits between and across biology, computer science, mathematics, statistics, software engineering, and other fields. Doing justice to these multiple and overlapping strands—networks, databases, software for manipulating sequences, and the hardware that they run on—means putting together an account that is "out of sequence." A combination of historical sources and ethnography (interviews and fieldwork) provides a series of cross sections through the recent history of the biosciences. Each of these cross sections shows, from a different perspective, how the flow of data has remade biological practice and knowledge.[22]

What do all these perspectives add up to in the end? They show that biology is in the midst of a fundamental transformation due to the introduction of computing. It is not adequate to characterize this shift as "digitization" or "informatization" or "dematerialization" or "liquification" of life.[23] Rather, we need to talk much more specifically about the ways in which computers have changed what count as interesting or answerable questions and satisfactory or validated solutions in biology. The material-to-virtual transition conceals a multiple and complex space of more subtle transitions: small to big, hypothesis-driven to data-driven, individuals to teams, and specific to general. Following the data allows us to observe the effects of an increasing fixation on sequence. The particular ease with which sequences (DNA, RNA, protein) can be treated as data—represented, stored, compared, and subjected to statistical analysis by computers—has both transformed them as objects and rendered them increasingly central in biological practice.[24] Our understanding of life, to an ever-greater extent, comes out of sequence.

But following the data also reveals ways in which sequence is arranged and manipulated. The "out of sequence" of the title can also be read in the more familiar sense as meaning "out of order" or "disordered." Bioinformatics reconstructs life not as a linear collection of genes, but as a densely connected network (of genes or proteins or organisms or species). The genome, we now see, is not in order; it is messy, mixed up. Data structures draw biological objects into new relationships, new topologies. Studying biology increasingly becomes about studying noise, randomness, and stochasticity. These are exactly the kinds of problems for which computers were first designed and used and which they are good at. Computers became widely used tools in biology only when these data-based modes of practice were imported from physics.

These three things—the ubiquity of computers and data in biology, the increasing importance of sequence, and stochastic approaches and understanding—are all linked. The use of computers in biology has provided epistemic space for particular sorts of problems—those based on large numbers, probability, and statistics. In so doing, it has also changed the objects that it studies—it has created new kinds of epistemic things. Sequences—and in particular the large sequence data sets from genomics—are particularly amenable to statistical analysis. It is the notion of the genome as a disordered object, or an object "out of sequence," that has allowed computers to become important tools for making sense of biology using statistical-computational methods. And, conversely, it is computers and computerization that have allowed sequences to continue to dominate the way we study and understand biology. The changing regimes of circulation and organization of biology are consequences of these epistemological shifts, rather than products of "informatization" per se.

In the end, "bioinformatics" is likely to disappear, to evaporate as a relevant term for describing biological practice. This will not happen because biologists stop using computers, but rather because the use of computers and bioinformatic methods will become so ubiquitous that it will make no sense to label it as a practice separate or distinct from biology generally. The kinds of knowledge and practices that are described here are already moving to the forefront of our understanding of life. It is crucial that we reflect on the consequences of this change—what difference does it make that we now examine, dissect, manipulate, and understand life with and through computers?

1 Building Computers

Before we can understand the effects of computers on biology, we need to understand what sorts of things computers *are*. Electronic computers were being used in biology even in the 1950s, but before 1980 they remained on the margins of biology—only a handful of biologists considered them important to their work. Now most biologists would find their work impossible without using a computer in some way. It seems obvious—to biologists as well as laypeople—that computers, databases, algorithms, and networks are appropriate tools for biological work. How and why did this change take place?

Perhaps it was computers that changed. As computers got better, a standard argument goes, they were able to handle more and more data and increasingly complex calculations, and they gradually became suitable for biological problems. This chapter argues that it was, in fact, the other way around: it was biology that changed to become a computerized and computerizable discipline. At the center of this change were data, especially sequence data. Computers are data processors: data storage, data management, and data analysis machines. During the 1980s, biologists began to produce large amounts of sequence data. These data needed to be collected, stored, maintained, and analyzed. Computers—data processing machines—provided a ready-made tool.

Our everyday familiarity with computers suggests that they are universal machines: we can use them to do the supermarket shopping, run a business, or watch a movie. But understanding the effects of computers—on biology at least—requires us to see these machines in a different light. The early history of computers suggests that they were not universal machines, but designed and adapted for particular kinds of data-driven problems. When computers came to be deployed in biology on a large scale, it was because these same kinds of problems became important in biology. Modes of thinking and working embedded in computational hardware were carried over from one discipline to another.

The use of computers in biology—at least since the 1980s—has entailed a shift toward problems involving statistics, probability, simulation, and stochastic methods. Using computers has meant focusing on the kinds of problems that computers are designed to solve. DNA, RNA, and protein sequences proved particularly amenable to these kinds of computations. The long strings of letters could be easily rendered as data and managed and manipulated as such. Sequences could be treated as patterns or codes that could be subjected to statistical and probabilistic analyses. They became objects ideally suited to the sorts of tools that computers offered. Bioinformatics is not just using computers to solve the same old biological problems; it marks a new way of thinking about and doing biology in which large volumes of data play the central role. Data-driven biology emerged because of the computer's history as a data instrument.

The first part of this chapter provides a history of early electronic computers and their applications to biological problems before the 1980s. It pays special attention to the purposes for which computers were built and the uses to which they were put: solving differential equations, stochastic problems, and data management. These problems influenced the design of the machines. Joseph November argues that between roughly 1955 and 1965, biology went from being an "exemplar of systems that computers could not describe to exemplars of systems that computers could indeed describe."[1] The introduction of computers into the life sciences borrowed heavily from operations research. It involved mathematizing aspects of biology in order to frame problems in modeling and data management terms—the terms that computers worked in.[2] Despite these adaptations, at the end of the 1970s, the computer still lay largely outside mainstream biological research. For the most part, it was an instrument ill-adapted to the practices and norms of the biological laboratory.[3]

The invention of DNA sequencing in the late 1970s did much to change both the direction of biological research and the relationship of biology with computing. Since the early 1980s, the amount of sequence data has continued to grow at an exponential rate. The computer was a perfect tool with which to cope with the overwhelming flow of data. The second and third parts of this chapter consist of two case studies: the first of Walter Goad, a physicist who turned his computational skills toward biology in the 1960s; and the second of James Ostell, a computationally minded PhD student in biology at Harvard University in the 1980s. These examples show how the practices of computer use were imported from physics into biology and struggled to establish themselves there. These practices became established as a distinct subdiscipline of biology—bioinformatics—during the 1990s.

What Is a Computer?

The computer was an object designed and constructed to solve particular sorts of problems, first for the military and, soon afterward, for Big Physics. Computers were (and are) good at solving certain types of problems: numerical simulations, differential equations, stochastic and statistical problems, and problems involving the management of large amounts of data.[4]

The modern electronic computer was born in World War II. Almost all the early attempts to build mechanical calculating devices were associated with weapons or the war effort. Paul Edwards argues that "for two decades, from the early 1940s until the early 1960s, the armed forces of the United States were the single most important driver of digital computer development."[5] Alan Turing's eponymous machine was conceived to solve a problem in pure mathematics, but its first physical realization at Bletchley Park was as a device to break German ciphers.[6] Howard Aiken's Mark I, built by IBM between 1937 and 1943, was used by the US Navy's Bureau of Ships to compute mathematical tables.[7] The computers designed at the Moore School of Electrical Engineering at the University of Pennsylvania in the late 1930s were purpose-built for ballistics computations at the Aberdeen Proving Ground in Maryland.[8] A large part of the design and the institutional impetus for the Electronic Numerical Integrator and Computer (ENIAC), also developed at the Moore School, came from John von Neumann. As part of the Manhattan Project, von Neumann was interested in using computers to solve problems in the mathematics of implosion. Although the ENIAC did not become functional until after the end of the war, its design—the kinds of problems it was supposed to solve—reflected wartime priorities.

With the emergence of the Cold War, military support for computers would continue to be of paramount importance. The first problem programmed onto the ENIAC (in November 1945) was a mathematical model of the hydrogen bomb.[9] As the conflict deepened, the military found uses for computers in aiming and operating weapons, weapons engineering, radar control, and the coordination of military operations. Computers like MIT's Whirlwind (1951) and SAGE (Semi-Automatic Ground Environment, 1959) were the first to be applied to what became known as C³I: command, control, communications, and intelligence.[10]

What implications did the military involvement have for computer design? Most early computers were designed to solve problems involving large sets of numbers. Firing tables are the most obvious example. Other problems, like implosion, also involved the numerical solution of differential equations.[11] A large set of numbers—representing an approximate solution—would be entered into the computer; a series of computations on these numbers would yield a new, better approximation. A solution could be approached iteratively. Problems such as radar control also involved (real-time) updating of large amounts of data fed in from remote military installations. Storing and iteratively updating large tables of data was the exemplary computational problem.

Another field that quickly took up the use of digital electronic computers was physics, particularly the disciplines of nuclear and particle physics. The military problems described above belonged strictly to the domain of physics. Differential equations and systems of linear algebraic equations can describe a wide range of physical phenomena such as fluid flow, diffusion, heat transfer, electromagnetic waves, and radioactive decay. In some cases, techniques of military computing were applied directly to physics problems. For instance, missile telemetry involved problems of real-time, multichannel communication that were also useful for controlling bubble chambers.[12] A few years later, other physicists realized that computers could be used to great effect in "logic" machines: spark chambers and wire chambers that used electrical detectors rather than photographs to capture subatomic events. Bubble chambers and spark chambers were complicated machines that required careful coordination and monitoring so that the best conditions for recording events could be maintained by the experimenters. By building computers into the detectors, physicists were able to retain real-time control over their experimental machines.[13]

But computers could be used for data reduction as well as control. From the early 1950s, computers were used to sort and analyze bubble chamber film and render the data into a useful form. One of the main problems for many particle physics experiments was the sorting of the signal from the noise: for many kinds of subatomic events, a certain "background" could be anticipated. Figuring out just how many background events should be expected inside the volume of a spark chamber was often a difficult problem that could not be solved analytically. Again following the lead of the military, physicists turned to simulations using computers. Starting with random numbers, physicists used stochastic methods that mimicked physical processes to arrive at "predictions" of the expected background. These "Monte Carlo" processes evolved from early computer simulations of atomic bombs on the ENIAC to sophisticated background calculations for bubble chambers. The computer itself became a particular kind of object: that is, a simulation machine.

The other significant use of computers that evolved between 1945 and 1955 was in the management of data. In many ways, this was a straightforward extension of the ENIAC's ability to work with large sets of numbers. The Moore School engineers J. Presper Eckert and John Mauchly quickly saw how their design for the Electronic Discrete Variable Advanced Calculator (EDVAC) could be adapted into a machine that could rapidly sort data—precisely the need of commercial work. This insight inspired the inventors to incorporate the Eckert-Mauchly Computer Corporation in December 1948 with the aim of selling electronic computers to businesses. The first computer they produced—the UNIVAC (Universal Automatic Computer)—was sold to the US Census Bureau in March 1951. By 1954, they had sold almost twenty machines to military (the US Air Force, US Army Map Service, Atomic Energy Commission) and nonmilitary customers (General Electric, US Steel, DuPont, Metropolitan Life, Consolidated Edison). Customers used these machines for inventory and logistics. The most important feature of the computer was its ability to "scan through a reel of tape, find the correct record or set of records, perform some process in it, and return the results again to tape."[14] It was an "automatic" information processing system. The UNIVAC was successful because it was able to store, operate on, and manipulate large tables of numbers—the only difference was that these numbers now represented inventory or revenue figures rather than purely mathematical expressions.

Between the end of World War II and the early 1960s, computers were also extensively used by the military in operations research (OR). OR and the related field of systems analysis were devoted to the systematic analysis of logistical problems in order to find optimally efficient solutions.[15] OR involved problems of game theory, probability, and statistics. These logical and numerical problems were understood as exactly the sorts of problems computers were good at solving.[16] The use of computers in OR and systems analysis not only continued to couple them to the military, but also continued their association with particular sorts of problems: namely, problems with large numbers of well-defined variables that would yield to numerical and logical calculations.[17]

What were the consequences of all this for the application of computers to biology? Despite their touted "universality," digital computers were not equally good at solving all problems. The ways in which early computers were used established standards and practices that influenced later uses.[18] The design of early computers placed certain constraints on where and how they would and could be applied to biological problems. The use of computers in biology was successful only where biological problems could be reduced to problems of data analysis and management. Bringing computers to the life sciences meant following specific patterns of use that were modeled on approaches in OR and physics and which reproduced modes of practice and patronage from those fields.[19]

In the late 1950s, there were two alternative notions of how computers might be applied to the life sciences. The first was that biology and biologists had to mathematize, becoming more like the physical sciences. The second was that computers could be used for accounting purposes, creating "a biology oriented toward the collation of statistical analysis of large volumes of quantitative data."[20] Both notions involved making biological problems amenable to computers' data processing power. Robert Ledley—one of the strongest advocates of the application of computers in biology and medicine—envisioned the transformation of biologists' research and practices along the lines of Big Science.[21]

In 1965, Ledley published *Use of Computers in Biology and Medicine*. The foreword (by Lee Lusted of the National Institutes of Health) acknowledged that computer use required large-scale funding and cooperation similar to that seen in physics.[22] Ledley echoed these views in his preface:

> Because of an increased emphasis on quantitative detail, elaborate experimentation and extensive quantitative data analysis are now required. Concomitant with this change, the view of the biologist as an individual scientist, personally carrying through each step of his data-reduction processes—that view is rapidly being broadened, to include the biologist as part of an intricate organizational chart that partitions scientific technical and administrative responsibilities.[23]

Physics served as the paradigm of such organization. But the physical sciences also provided the model for the kinds of problems that computers were supposed to solve: those involving "large masses of data and many complicated interrelating factors." Many of the biomedical applications of computers that Ledley's volume explored treated biological systems according to their physical and chemical bases. The examples Ledley describes in his introduction include the numerical solution of differential equations describing biological systems (including protein structures, nerve fiber conduction, muscle fiber excitability, diffusion through semipermeable membranes, metabolic reactions, blood flow), simulations (Monte Carlo simulation of chemical reactions, enzyme systems, cell division, genetics, self-organizing neural nets), statistical analyses (medical records, experimental data, evaluation of new drugs, data from electrocardiograms and electroencephalograms, photomicrographic analysis); real-time experimental and clinical control (automatic respirators, analysis of electrophoresis, diffusion, and ultracentrifuge patterns, and counting of bacterial cultures) and medical diagnosis (including medical records and distribution and communication of medical knowledge).[24] Almost all the applications were either borrowed directly from the physical sciences or depended on problems involving statistics or large volumes of information.[25]

For the most part, the mathematization and rationalization of biology that Ledley and others believed was necessary for the "computerization" of the life sciences did not eventuate.[26] By the late 1960s, however, the invention of minicomputers and the general reduction in the costs of computers allowed more biologists to experiment with their use.[27] At Stanford University, a small group of computer scientists and biologists led by Edward Feigenbaum and Joshua Lederberg began to take advantage of these changes. After applying computers to the problem of determining the structure of organic molecules, this group began to extend their work into molecular biology.[28]

In 1975, they created MOLGEN, or "Applications of Symbolic Computation and Artificial Intelligence to Molecular Biology." The aim of this project was to combine expertise in molecular biology with techniques from artificial intelligence to create "automated methods for experimental assistance," including the design of complicated experimental plans and the analysis of nucleic acid sequences.[29] Lederberg and Feigenbaum initially conceived MOLGEN as an artificial intelligence (AI) project for molecular biology. MOLGEN included a "knowledge base" compiled by expert molecular biologists and containing "declarative and procedural information about structures, laboratory conditions, [and] laboratory techniques."[30] They hoped that MOLGEN, once provided with sufficient information, would be able to emulate the reasoning processes of a working molecular biologist. Biologists did not readily take up these AI tools, and their use remained limited. What did begin to catch on, however, were the simple tools created as part of the MOLGEN project for entering, editing, comparing, and analyzing protein and nucleic acid sequences. In other words, biologists used MOLGEN for data management, rather than for the more complex tasks for which it was intended.

By the end of the 1970s, computers had not yet exerted a wide influence on the knowledge and practice of biology. Since about 1975, however, computers have changed what it means to do biology: they have "computerized" the biologist's laboratory. By the early 1980s, and especially after the advent of the first personal computers, biologists began to use computers in a variety of ways. These applications included the collection, display, and analysis of data (e.g., electron micrographs, gel electrophoresis), simulations of molecular dynamics (e.g., binding of enzymes), simulations of evolution, and especially the study of the structure and folding of proteins (reconstructing data from X-ray crystallography, visualization, simulation and prediction of folding).[31] However, biologists saw the greatest potential of computers in dealing with sequences. In 1984, for instance, Martin Bishop wrote a review of software for molecular biology; out of fifty-three packages listed, thirty were for sequence analysis, a further nine for "recombinant DNA strategy," and another seven for database retrieval and management.[32] The analysis of sequence data was becoming the exemplar for computing in biology.[33]

As data processing machines, computers could be used in biology only in ways that aligned with their uses in the military and in physics. The early design and use of computers influenced the ways in which they could and would be applied in the life sciences. In the 1970s, the computer began to bring new kinds of problems (and techniques for solving them) to the fore in biology—simulation, statistics, and large-volume data management and analysis were the problems computers could solve quickly. We will see how these methods had to struggle to establish themselves within and alongside more familiar ways of knowing and doing in the life sciences.

Walter Goad and the Origins of GenBank

The next two sections provide two examples of ultimately successful attempts to introduce computers into biology. What these case studies

suggest is that success depended not on adapting the computer to biological problems, but on adapting biology to problems that computers
could readily solve. In particular, they demonstrate the central roles that
data management, statistics, and sequences came to play in these new
kinds of computationally driven biology. Together, these case studies
also show that the application of computers to biology was not obvious
or straightforward—Goad was able to use computers only because of
his special position at Los Alamos, while Ostell had to struggle for many
years to show the relevance and importance of his work. Ultimately,
the acceptance of computers by biologists required a redefinition of the
kinds of problems that biology addressed.

Walter Goad (1925–2000) came to Los Alamos Scientific Laboratories as a graduate student in 1951, in the midst of President Truman's crash program to construct a hydrogen bomb. He quickly proved
himself an able contributor to that project, gaining key insights into
problems of neutron flux inside supercritical uranium. There is a clear
continuity between some of Goad's earlier (physics) and later (biological) work: both used numerical and statistical methods to solve data-
intensive problems. Digital electronic computers were Goad's most important tool. As a consequence, Goad's work imported specific ways of
doing and thinking from physics into biology. In particular, he brought
ways of using computers as data management machines. Goad's position as a senior scientist in one of the United States' most prestigious scientific research institutions imparted a special prestige to these modes of
practice. Ultimately, the physics-born computing that Goad introduced
played a crucial role in redefining the types of problems that biologists
addressed; the reorganization of biology that has accompanied the genomic era can be understood in part as a consequence of the modes of
thinking and doing that the computer carried from Los Alamos.

We can reconstruct an idea of the kinds of physics problems that
Goad was tackling by examining both some of his published work from
the 1950s and his thesis on cosmic ray scattering.[34] This work had three
crucial features. First, it depended on modeling systems (like neutrons)
as fluids using differential or difference equations. Second, such systems
involved many particles, so their properties could only be treated statistically. Third, insight was gained from the models by using numerical or
statistical methods, often with the help of a digital electronic computer.
During the 1950s, Los Alamos scientists pioneered new ways of problem solving using these machines.

Electronic computers were not available when Goad first came to
Los Alamos in 1951 (although Los Alamos had had access to comput-

ers elsewhere since the war). By 1952, however, the laboratory had the MANIAC (Mathematical Analyzer, Numerical Integrator, and Computer), which had been constructed under the direction of Nicholas Metropolis. Between 1952 and 1954, Metropolis worked with Enrico Fermi, Stanislaw Ulam, George Gamow, and others on refining Monte Carlo and other numerical methods for use on the new machine. They applied these methods to problems in phase-shift analysis, nonlinear-coupled oscillators, two-dimensional hydrodynamics, and nuclear cascades.[35] Los Alamos also played a crucial role in convincing IBM to turn its efforts to manufacturing digital computers in the early 1950s. It was the first institution to receive IBM's "Defense Calculator," the IBM 701, in March 1953.[36]

When attempting to understand the motion of neutrons inside a hydrogen bomb, it is not possible to write down (let alone solve) the equations of motion for all the neutrons (there are far too many). Instead, it is necessary to find ways of summarizing the vast amounts of data contained in the system. Goad played a central role in Los Alamos' work on this problem. By treating the motion of neutrons like the flow of a fluid, Goad could describe it using well-known differential equations. These equations could be solved by "numerical methods"—that is, by finding approximate solutions through intensive calculation.[37] In other cases, Goad worked by using Monte Carlo methods—that is, by simulating the motion of neutrons as a series of random moves.[38] In this kind of work, Goad used electronic computers to perform the calculations: the computer acted to keep track of and manage the vast amounts of data involved. The important result was not the motion of any given neutron, but the overall pattern of motion, as determined from the statistical properties of the system.

When Goad returned to his thesis at the end of 1952, his work on cosmic rays proceeded similarly. He was attempting to produce a model of how cosmic rays would propagate through the atmosphere. Since a shower of cosmic rays involved many particles, once again it was not possible to track all of them individually. Instead, Goad attempted to develop a set of equations that would yield the statistical distribution of particles in the shower in space and time. These equations were solved numerically based on theoretical predictions about the production of mesons in the upper atmosphere.[39] In both his work on the hydrogen bomb and his thesis, Goad's theoretical contributions centered on using numerical methods to understand the statistics of transport and flow.

By the 1960s, Goad had become increasingly interested in some problems in biology. While visiting the University of Colorado Medical

Center, Goad collaborated extensively with the physical chemist John R. Cann, examining transport processes in biological systems. First with electrophoresis gels, and then extending their work to ultracentrifugation, chromatography, and gel filtration, Goad and Cann developed models for understanding how biological molecules moved through complex environments.[40] The general approach to such problems was to write down a set of differential or difference equations that could then be solved using numerical methods on a computer. This work was done on the IBM-704 and IBM-7094 machines at Los Alamos. These kinds of transport problems are remarkably similar to the kinds of physics that Goad had contributed to the hydrogen bomb: instead of neutrons moving through a supercritical plasma, the equations now had to represent macromolecules moving through a space filled with other molecules.[41]

Here too, it was not the motion of any particular molecule that was of interest, but the statistical or average motion of an ensemble of molecules. Such work often proceeded by treating the motion of the molecule as a random walk and then simulating the overall motion computationally using Monte Carlo methods. Goad himself saw some clear continuities between his work in physics and his work in biology. Reporting his professional interests in 1974, he wrote, "Statistics and statistical mechanics, transport processes, and fluid mechanics, especially as applied to biological and chemical phenomena."[42] By 1974, Goad was devoting most of his time to biological problems, but "statistical mechanics, transport processes, and fluid mechanics" well described his work in theoretical physics too. Likewise, in a Los Alamos memo from 1972, Goad argued that the work in biology should not be split off from the Theoretical Division's other activities: "Nearly all of the problems that engage [Los Alamos] have a common core: . . . the focus is on the behavior of macroelements of the system, the behavior of microelements being averaged over—as in an equation of state—or otherwise statistically characterized."[43] Goad's work may have dealt with proteins instead of nucleons, but his modes of thinking and working were very similar. His biology drew on familiar tools, particularly the computer, to solve problems by deducing the statistical properties of complex systems. The computer was the vital tool here because it could keep track of and summarize the vast amounts of data present in these models.

Los Alamos provided a uniquely suitable context for this work. The laboratory's long-standing interest in biology and medicine—and particularly in molecular genetics—provided some context for Goad's forays. Few biologists were trained in the quantitative, statistical, and numerical methods that Goad could deploy; even fewer had access to expensive, powerful computers. Mathematical biology remained an extremely isolated subdiscipline.[44] Those few who used computers for biology were marginalized as theoreticians in an experiment-dominated discipline. Margaret Dayhoff at the National Biomedical Research Foundation (NBRF), for instance, struggled to gain acceptance among the wider biological community.[45] Goad's position at Los Alamos was such that he did not require the plaudits of biologists—the prestige of the laboratory itself, as well as its open-ended mission, allowed him the freedom to pursue a novel kind of cross-disciplinary work.

In 1974, the Theoretical Division's commitment to the life sciences was formalized by the formation of a new subdivision: T-10, Theoretical Biology and Biophysics. The group was formally headed by George I. Bell, but by this time Goad too was devoting almost all his time to biological problems. The group worked on problems in immunology, radiation damage to nucleotides, transport of macromolecules, and human genetics. The senior scientists saw their role as complementary to that of the experimenters, building and analyzing mathematical models of biological systems that could then be tested.[46]

It was around this time that Goad and the small group of physicists working with him began to devote more attention to nucleic acid sequences. For biologists, both protein and nucleotide sequences were the keys to understanding evolution. Just as morphologists compared the shapes of the bones or limbs of different species, comparing the sequence of dog hemoglobin with that of cow hemoglobin, for instance, allowed inferences about the relatedness and evolutionary trajectories of dogs and cows. The more sequence that was available, and the more sensitively it could be compared, the greater insight into evolution could be gained. In other words, sequence comparison allowed biologists to study evolutionary dynamics very precisely at the molecular level.[47] Sequences were an appealing subject of research for physicists for several reasons. First, they were understood to be the fundamental building blocks of biology—studying their structure and function was equivalent in some sense to studying electrons and quarks in physics. Second, their discrete code seemed susceptible to the quantitative and computational tools that physicists had at their disposal. Computers were useful for processing the large quantities of numerical data from physics experiments and simulations; the growth of nucleotide sequence data offered similar possibilities for deploying the computer in biology.[48] The T-10 group immediately attempted to formulate sequence analysis as a set of mathematical problems. Ulam realized quickly that the problem of comparing sequences with one another was really a problem of finding a "metric space of sequences."[49]

Under the supervision of Goad, a group of young physicists—including Temple Smith, Michael Waterman, Myron Stein, William A. Beyer, and Minoru Kanehisa—began to work on these problems of sequence comparison and analysis, making important advances both mathematically and in software.[50] T-10 fostered a culture of intense intellectual activity; its members realized that they were pursuing a unique approach to biology with skills and resources available to few others.[51] Within the group, sequence analysis was considered a problem of pattern matching and detection: within the confusing blur of As, Gs, Ts, and Cs in a DNA sequence, lay hidden patterns that coded for genes or acted as protein-specific binding sites. Even the relatively short (by contemporary standards) nucleotide sequences available in the mid-1970s contained hundreds of base pairs—far more than could be made sense of by eye. As a tool for dealing with large amounts of data and for performing statistical analysis, the computer was ideal for sequence analysis.[52] Goad's earlier work in physics and biology had used computers to search for statistical patterns in the motion of neutrons or macromolecules; here, also by keeping track of large amounts of data, computerized stochastic techniques (e.g., Monte Carlo methods) could be used for finding statistical patterns hidden in the sequences. As the *Los Alamos News Bulletin* said of Goad's work on DNA in 1982, "Pattern-recognition research and the preparation of computer systems and codes to simplify the process are part of a long-standing effort at Los Alamos—in part the progeny of the weapons development program here."[53] Goad's work used many of the same tools and techniques that had been developed at the laboratory since its beginnings, applying them now to biology instead of bombs.

The Los Alamos Sequence Database—and eventually GenBank—evolved from these computational efforts. For Goad, the collection of nucleotide sequences went hand in hand with their analysis: collection was necessary in order to have the richest possible resource for analytical work, but without continuously evolving analytical tools, a collection would be just a useless jumble of base pairs. In 1979, Goad began a pilot project with the aim of collecting, storing, analyzing, and distributing nucleic acid sequences. This databasing effort was almost coextensive with the analytical work of the T-10 group: both involved using the computer for organizing large sets of data. The techniques of large-scale data analysis required for sequence comparison were very similar to the methods required for tracking and organizing sequence in a database. "These activities have in common," Bell wrote, "enhancing our understanding of the burgeoning data of molecular genetics both by relatively straightforward organization and analysis of the data and by the development of new tools for recognizing important features of the data."[54] Databasing meant knowing how to use a computer for organizing and keeping track of large volumes of data. In other words, data management—the organization of sequence data into a bank—depended deeply on the kinds of computer-based approaches that Goad had been using for decades in both physics and biology. Goad's experience with computers led him (and the T-10 group) to understand and frame biological problems in terms of pattern matching and data management—these were problems that they possessed the tools to solve. In so doing, these physicists brought not only new tools to biology, but new kinds of problems and practices.

In 1979, Goad submitted an unsolicited proposal to the National Institutes of Health (NIH) in the hope that he might receive funding to expand his database. After some hesitation, a competitive request for proposals was issued by the NIH in 1981. Goad was not the only person attempting to collect sequences and organize them using computers. Elvin Kabat had begun a collection of sequences of immunoglobulins at the NIH, while Kurt Stüber in Germany, Richard Grantham in France, and Douglas Brutlag at Stanford also had their own sequence collections. Dayhoff used computer analysis to compile the *Atlas of Protein Sequence and Structure* from her collection of (mostly protein) sequences at the NBRF. However, this kind of collection and analysis was not considered high-prestige work by biologists, and Dayhoff struggled to find funding for her work.[55] Goad's position as a physicist at a prestigious laboratory afforded him independence from such concerns: he could pursue sequence collection and comparison just because he thought it was valuable scientific work.

Ultimately, the $2 million, five-year contract for the publicly funded sequence database was awarded to Goad's group in June 1982. Both the origins and the subsequent success of GenBank have been detailed elsewhere.[56] Goad's scientific biography, however, suggests that GenBank was partly a product of his background in physics, as he imported a statistical and data management style of science into biology via the computer. Goad's position as a physicist at a world-renowned laboratory allowed him to import ways of working into biology from his own discipline. Goad's techniques and tools—particularly the computer—carried their prestige from his work in physics and had a credibility that did not depend on norms of biological work. These circumstances allowed the introduction not only of a new tool (the computer), but also of specific ways of thinking centered on statistics, pattern recognition, and data management. Goad's background meant that the computer came to biology not as a machine for solving biological problems, but rather as a technology that imported ready-made ways of thinking, doing, and organizing from physics.

From Sequence to Software: James Ostell

It is important not to exaggerate the extent of computer use in molecular biology in the early 1980s. One MOLGEN report from September 1980 provides a list of just fifty-one users who had logged into the system.[57] Although interest and use were growing rapidly, computers still remained esoteric tools for most molecular biologists. This certainly appeared to be true for Jim Ostell when he began his doctoral studies in the laboratory of Fotis Kafatos in Harvard's Department of Cellular and Developmental Biology in 1979.[58] Although Ostell had a background in zoology (he wrote a master's thesis on the anatomy of the male cricket at the University of Massachusetts), he was attracted to the exciting field of molecular biology, which seemed to be advancing rapidly due to the new techniques of DNA sequencing and cDNA cloning. Swept up in the excitement, Ostell did some cloning and sequencing of eggshell proteins. Once he had the sequence, however, he had no idea what to do next, or how to make any sense of it. Somebody suggested that he use a computer. Before coming to graduate school, Ostell had taken one computer class, using the FORTRAN programming language on a Cyber 70 mainframe with punched cards.[59] The Kafatos lab had a 300-baud modem that connected an ASCII terminal to the MOLGEN project running at Stanford. It also had an 8-bit CP/M microcomputer with an Intel CPU, 48 kilobytes of memory, and an 8-inch floppy disk drive.[60] The secretary had priority for the use of the computer for word processing, but the students were free to use it after hours. Ostell relates how he came to use the machine:

> This computer was always breaking down, so the repair people were often there. I had been a ham radio operator and interested in electronics, so Fotis [Kafatos] found me one day looking interestedly in the top as it was under repair and asked if I knew anything about computers. When I replied "A little," he smiled and said "Great! You are in charge of the computer."[61]

Ostell had begun to experiment with the MOLGEN system for analyzing his sequences. He found the tools it provided unsatisfactory for
his purposes. As a result, he began to write his own sequence analysis
software in FORTRAN, using a compiler that had apparently come with
the computer. In his dissertation, written some seven years later, Ostell
outlined two sorts of differences between the MOLGEN programs and
his own. First, the MOLGEN software was designed to run on a mainframe, "supported by substantial government grants." By contrast, the
system that Ostell was using was "mainly the province of computer
buffs and cost about $10,000. . . . It was a radical idea that comparable
performance could be attained from a (relatively) inexpensive desktop
computer."[62] Second, Ostell's programs were user-friendly:

> The software attempted to converse with the scientist in an
> immediately understandable way. Instead of questions like
> ">MAXDIST?_," such as one would encounter on the Molgen
> system, this package would ask things like "What is the maxi-
> mum distance to analyze (in base pairs)?" The other aspect of
> "doing biology" was the way the analyses were done. For exam-
> ple, the Molgen software would give the positions of restriction
> enzyme recognition sites in a sequence. But why would a biolo-
> gist want to do a restriction search in the first place? Probably
> to plan a real experiment. So my package would give the cut site
> for the enzyme, not the recognition site. . . . I feel mine provided
> more immediately useful information to the scientist.[63]

Ostell's colleagues, first in his own lab, then all over the Harvard Bio-
Labs, soon began asking to use his programs. When he published a
description of the programs in *Nucleic Acids Research* in 1982, offering free copies to anyone who wanted it, he was overwhelmed with
requests.[64] Ostell's programs constituted one of the most complete
software packages available for molecular biology and the only one
that would function on a microcomputer. In addition to making his
programs microcomputer-friendly, Ostell made sure that they could be
compiled and used on multiple platforms. Roger Staden's similar package suffered from the fact that it used unusual FORTRAN commands
and made occasional PDP-11 system calls (that is, it was designed for
a mainframe).[65] Over the next few years, Ostell occupied himself with
making the package available to as wide a range of collaborators as possible, adapting it for different systems and adding additional features.

In a description of an updated version of his programs published in 1984, Ostell made a bold claim for the value of his work:

> Adequate understanding of the extensive DNA and protein sequence derived by current techniques requires the use of computers. Thus, properly designed sequence analysis programs are as important to the molecular biologist as are experimental techniques.[66]

Not everyone shared his view, including some members of Ostell's PhD committee at Harvard. "It wasn't something that biologists should be doing," according to Ostell's recollection of the reaction of some members of his committee.[67] Despite Kafatos's support, Ostell was not permitted to graduate on the basis of his computational work. Ostell's programs had made a direct contribution to the solution of many biological problems, but the software itself was not understood to be "doing biology." Even in Ostell's own writing about his work, he describes the functions of his programs as "data management" and "data analysis," rather than biology proper.[68]

Ostell could not get his PhD, but Kafatos agreed to allow him to stay on as a graduate student provided that he could support himself. He got permission to teach a class called "Computer Anatomy and Physiology" to undergraduates. This class was an introduction to computer hardware that analyzed the machine as if it were a living organism. In 1984, Ostell was approached by International Biotechnologies, Inc. (IBI, a company that had been selling restriction enzymes and laboratory equipment), which wanted to license his software and develop it into a product. Since Ostell had done the work while a graduate student at Harvard, the university had a legal claim to the intellectual property rights. But it saw no commercial value in Ostell's work and agreed to sign over all rights. Turning the software into a commercial product was a formidable task. In particular, the software had to be carefully reengineered for reliability, compatibility, and interoperability. The IBI/Pustell Sequence Analysis Package was released in August 1984, ready to use on an IBM personal computer, at a cost of $800 for academic and commercial users. Still unable to graduate, Ostell followed his wife's medical career to Vermont, where he lived in a nineteenth-century farmhouse and adapted his programs for use, first on MS-DOS and Unix machines and then on the new Apple Macintosh computers (the latter version eventually became the MacVector software).

In an attempt to convince his committee and others of the value of his work, Ostell also embarked on applying his software to various biological problems, collaborating with others in the Harvard BioLabs. This effort resulted in significant success, particularly in using his programs to analyze conservation patterns and codon bias to determine protein-coding regions and exon boundaries in *Drosophila* and broad bean (*Vicia faba*) genes.[69] These sorts of problems have two significant features. First, they require the manipulation and management of large amounts of data. Analysis of conservation patterns, for instance, requires organizing sequences from many organisms according to homology before performing comparisons. Second, analyzing codon bias and finding protein-coding regions are statistical problems. They treat sequences as a stochastic space, where the problem is one of finding a "signal" (a protein-coding region) amid the "noise" of bases. Consider this excerpt from Ostell's thesis in which he explains how codon bias is calculated:

> Each sequence used to make the table is then compared to every other sequence in the table by Pearson product moment correlation coefficient. This is, the bias is calculated for each codon in each of two sequences being compared. The correlation coefficient is then calculated comparing the bias for every codon between the two sequences. The correlation coefficient gives a sense of the "goodness of fit" between the two tables. A correlation coefficient is also calculated between the sequence and aggregate table. Finally a C statistic, with and without strand adjustment, is calculated for the sequence on both its correct and incorrect strands. These calculations give an idea how well every sequence fits the aggregate data, as well as revealing relationships between pairs of sequences.[70]

Countless similar examples could be taken from the text: the basis of Ostell's programs was the use of the computer as a tool for managing and performing statistical analysis on sequences.

The story has a happy ending. By 1987, Ostell's committee allowed him to submit his thesis. As David Lipman began to assemble a team for the new National Center for Biotechnology Information (NCBI), he realized that he had to employ Ostell—there was no one else in the world who had such a deep understanding of the informational needs of biologists. Selling the rights to his software so as not to create a conflict of interest, Ostell began work at the NCBI in November 1988 as the chief of information engineering (a position in which he remains in 2012).[71] In particular, Lipman must have been impressed by Ostell's vision for integrating and standardizing biological information. As Ostell's work and thinking evolved, it became clear to him that the way data were stored and managed had fundamental significance for biological practice and knowledge. Developing a new set of tools that he called a "cyborg software environment," Ostell attempted to allow the user to interface directly with the sequence, placing the DNA molecule at the center of his representation of biological information.

> Computer images of DNA sequences have been strongly influenced by this vision of an isolated object. We sequence a piece of DNA and read a series of bases as a linear series of bands on a gel. On paper we represent DNA as a linear series of letters across a page. Virtually every computer program which operates on DNA sequences represents a DNA sequence as a linear series of bytes in memory, just as its representation on a printed page. However, a typical publication which contains such a linear series of letters describing a particular DNA always contains much more information. . . . Most computer programs do not include any of this information.[72]

Ostell proposed a new way of representing this extra information that "tie[d] all annotations to the simple coordinate system of the sequence itself."[73] The computer now provided a way to order biology and think about biological problems in which sequences played the central role.

By 1987, as he was finishing his thesis, Ostell realized that he had been involved in "the beginnings of a scientific field."[74] Problems that had been impossibly difficult in 1980 were now being solved as a matter of routine. More importantly, the problems had changed: whereas Ostell had begun by building individual programs to analyze particular sequences, by the late 1980s he was tackling the design of "software environments" that allowed integrated development of tools and large-scale data sharing that would transcend particular machines and file formats. For the epigraph to the introduction to his thesis, Ostell quoted Einstein: "Opinions about obviousness are to a certain extent a function of time." Given the difficulties Ostell faced in completing his degree, it is hard not to read this quotation as a comment on the discipline he had helped to create: the application of computers to biology had gone from the "unobvious" to the "obvious."

But why? What does Ostell's story suggest about the transition?

First, it makes clear the importance of sequence and sequencing: the growth of sequence created a glut of data that had to be managed. The computerization of biology was closely associated with the proliferation of sequences; sequences were the kinds of objects that could be manipulated and interrogated using computers. Second, the growing importance of sequences increased the need for data management. The design of computers was suited to knowledge making through the management and analysis of large data sets.

Disciplinary Origins of Bioinformatics

In the 1980s, due to the perseverance of Ostell and others like him, computers began to become more prevalent in biological work. The individuals who used these machines brought with them commitments to statistical and data management approaches originating in physics. During the 1990s, the use of computers in biology grew into a recognized and specialized set of skills for managing and analyzing large volumes of data.

The *Oxford English Dictionary* now attributes the first use of the term "bioinformatics" to Paulien Hogeweg in 1978, but there is a strong case to be made that the discipline, as a recognized subfield of biology, did not come into existence until the early 1990s.[75] Searching PubMed—a comprehensive online citation database for the biomedical sciences—for the keyword "bioinformatics" from 1982 to 2008 suggests that the field did not grow substantially until about 1992 (figure 1.1). Although this is not a perfect indicator, it suffices to show the general trend.[76] From a trickle of publications in the late 1980s, the early 1990s saw an increase to several hundred papers per year (about one paper per day). This number remained relatively steady from 1992 to 1998, when the field underwent another period of rapid growth, up to about ten thousand papers per year (about twenty-seven papers per day) in 2005.

The late 1980s and early 1990s also saw the founding of several key institutions, including the NCBI in 1988 and the European Bioinformatics Institute in 1993. In 1990 and 1991, the Spring Symposia on Artificial Intelligence and Molecular Biology were held at Stanford.[77] Lawrence E. Hunter, a programmer at the National Library of Medicine (NLM), was one of the organizers: "It was really hard to find people who did this work in either computer science or molecular biology. No one cared about bioinformatics or had any idea of what it was or how to find people who did it."[78] In 1992, Hunter used publications

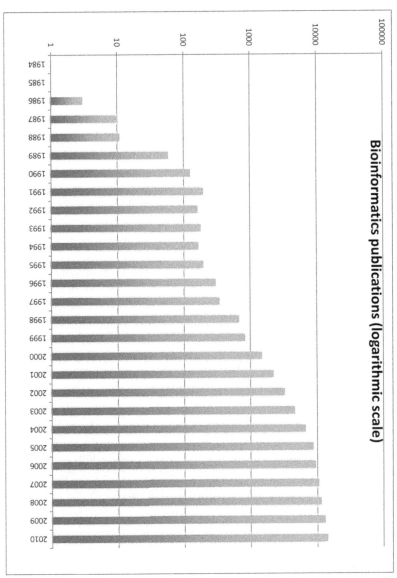

FIGURE 1.1 Number of publications in PubMed with keywords "bioinformatics," "computational biology," and "computational AND biology" between 1984 and 2010 (logarithmic scale). Two periods of rapid growth can be distinguished between 1988 and 1991 and between 1998 and 2004. (Compiled by author from PubMed, August 19, 2011.)

and conference mailing lists to generate a database of researchers interested in artificial intelligence and molecular biology. The conference that Hunter organized around his list became the first Intelligent Systems for Molecular Biology (ISMB) meeting, held in Washington, DC, in 1993 and jointly sponsored by the NLM and the National Science Foundation.[79]

It was also during the early 1990s that the first moves were made toward establishing specially designed bioinformatics courses at the undergraduate and graduate levels. In 1993, undergraduate and doctoral programs were established in bioinformatics at Baylor College of Medicine, Rice University, and the University of Houston. These were followed in 1996 by programs at Northwestern University, Rutgers University, and the University of Washington. By 1999, according to one report, there were twenty-one bioinformatics programs in the United States.[80]

What caused this institutionalization of bioinformatics? It would be possible to tell this story as part of a history of the Human Genome Project (HGP)—many of the sequence data on which computers went to work were generated as part of human genome mapping and sequencing efforts. However, the story might just as easily be told the other way around: the HGP became a plausible and thinkable project only because methods of managing large amounts of data were already coming into existence in the 1980s. Computers had begun to be used for managing sequence data before the HGP's beginnings. The HGP, while crystallizing the institutional development of bioinformatics, depended on the prior existence of bioinformatic practices of data management. It was the possibility of being able to store and manage the 3 billion base pairs of the human genome on a computer that made the project make sense. Similar problems of data management had arisen *before* the HGP.

Almost since GenBank's inception in 1982, its managers and advisors were under constant pressure to keep up with the increases in the publication of sequence data. By 1985, Los Alamos (along with Bolt, Beranek and Newman, which shared the responsibility for running GenBank) was complaining of the difficulties of keeping up with the rate of sequence production: "The average amount of new information [each month] . . . is fully half the size of the first GenBank release in October 1982."[81] The cause of their concern was clear: a fixed budget and a fixed number of staff had to cope with an exponentially growing set of sequences. Coping with this problem was the major challenge for Gen-Bank throughout the 1980s—the need for timely entry and completeness of data constantly animated GenBank staff, its advisory board, and its NIH overseers.

But the problem extended beyond a single database. In 1985, an editorial in the first issue of the journal *Computer Applications in the Biosciences: CABIOS* warned, "The information 'explosion' cannot continue indefinitely, but has already reached unmanageable proportions."[82] An announcement for a seminar in October 1987 expressed a similarly dire sentiment: "The body of experimental data in the biological sciences is immense and growing rapidly. Its volume is so extensive that computer methods, possibly straining the limits of current technology will be necessary to organize the data."[83] By the time the HGP had begun in earnest, the problem was even worse: "Data collection is outstripping current capabilities to annotate, store, retrieve, and analyze maps and sequences."[84] The amounts of data seemed to be constantly overwhelming biologists' abilities to analyze and understand them. However, computers seemed to present a ready-made solution: they were designed to handle exactly the kinds of data management problems that biology now presented. Their roots in Big Science made them suitable tools for controlling the growing number of base pairs.

In describing their crisis, biologists often used the metaphor of a "data flood." Biological data, like water, are generally a good thing, necessary for the growth of biological understanding. But in large quantities (and moving at high speed), data, like water, can be dangerous—they can submerge the structures on which knowledge is built.[85] Of particular concern to biologists was the fact that as molecular biological data were generated, many of them flowed in an ad hoc fashion—into isolated databases, in nonstandard formats. Already by the early 1990s, finding all the available information about a particular gene or sequence was becoming nearly impossible. Using the computer as a day-to-day laboratory tool was one thing, but organizing large-scale coordination and collaboration required specialized knowledge and skills. The diluvian metaphor contributed to the sense that controlling the flow of data, filtering them and channeling them to the appropriate places, was an activity distinct from, but just as important as, the eventual use of the data. In other words, the metaphor created an epistemic space for data management in biology.

By the time the HGP came on the scene, databases such as GenBank had already turned to computers and bioinformatic techniques for data management. Efforts to create institutional spaces for data work also arose independently of the HGP.[86] The bill calling for the creation of

NCBI, first brought before Congress in 1986, reasoned that "knowledge in the field of biotechnology is accumulating faster than it can reasonably be assimilated" and that advances in computation were the solution.[87] In particular, the bill recognized that the design, development, implementation, and management of biological information constituted a set of distinct skills that required a distinct institutional home. Likewise, in Europe, the formation of the European Bioinformatics Institute (EBI) as a quasi-independent outstation of the European Molecular Biology Laboratory (EMBL) was justified on the grounds that the application of computing tools to biology constituted a necessary and distinct skill set.

> Sophisticated interacting information resources must be built both from EMBL Data Library products and in collaboration with groups throughout Europe. Support and training in the use and development of such resources must be provided. Research necessary to keep them state-of-the-art must be carried out. Close links with all constituents must be maintained. These include research scientists, biotechnologists, software and computer vendors, scientific publishers, and government agencies. . . . Increased understanding of biological processes at the molecular level and the powerful technologies of computer and information science will combine to allow bioinformatics to transcend its hitherto largely service role and make fundamentally innovative contributions to research and technology.[88]

The computer was no longer to be considered a mere lab tool, but rather the basis of a discipline that could make "innovative contributions" to biomedicine.

The knowledge and skills that were needed to manage this data crisis were those that been associated with computing and computers since the 1950s. In particular, molecular biology needed data management and statistics. From the mid-1990s onward, new journals, new conferences, and new training programs (and textbooks) began to appear to support this new domain of knowledge. In 1994, the new *Journal of Computational Biology* announced that

> computational biology is emerging as a discipline in its own right, in much the same way molecular biology did in the late 1950s and early 1960s. . . . Biology, regardless of the sub-specialty, is overwhelmed with large amounts of very complex data. . . .

Thus all areas of the biological sciences have urgent needs for the organized and accessible storage of biological data, powerful tools for analysis of those data, and robust, mathematically based, data models. Collaborations between computer scientists and biologists are necessary to develop information platforms that accommodate the need for variation in the representation of biological data, the distributed nature of the data acquisition system, the variable demands placed on different data sets, and the absence of adequate algorithms for data comparison, which forms the basis of biological science.[89]

In 1995, Michael Waterman published *Introduction to Computational Biology: Maps, Sequences, and Genomes*, arguably the first bioinformatics textbook.[90] Waterman's book was grounded in the treatment of biology as a set of statistical problems: sequence alignment, database searching, genome mapping, sequence assembly, RNA secondary structure, and evolution, are all treated as statistical problems.[91]

In 1998, the fourteen-year-old journal *Computer Applications in the Biosciences* changed its name to *Bioinformatics*. In one of the first issues, Russ Altman outlined the need for a specialized curriculum for bioinformatics. Altman stressed that "bioinformatics is not simply a proper subset of biology or computer science, but has a growing and independent base of tenets that requires specific training not appropriate for either biology or computer science alone."[92] Apart from the obvious foundational courses in molecular biology and computer science, Altman recommended training for future bioinformaticians in statistics (including probability theory, experimental statistical design and analysis, and stochastic processes) as well as several specialized domains of computer science: optimization (expectation maximization, Monte Carlo, simulated annealing, gradient-based methods), dynamic programming, bounded search algorithms, cluster analysis, classification, neural networks, genetic algorithms, Bayesian inference, and stochastic context-free grammars.[93] Almost all of these methods trace their origins to statistics or physics or both.[94]

Altman's editorial was inspired by the fact that solid training in bioinformatics was hard to come by. The kinds of skills he pointed to were in demand, and bioinformatics was the next hot career. Under the headline "Bioinformatics: Jobs Galore," *Science Careers* reported in 2000 that "everyone is struggling to find people with the bioinformatics skills they need."[95] The period 1997–2004 coincides with the second period of rapid growth of bioinformatics publications (see figure 1.1). Between

1999 and 2004, the number of universities offering academic programs in bioinformatics more than tripled, from 21 to 74.[96] The acceleration and rapid completion of the HGP made it clear that, as one *Nature* editor put it, "like it or not, big biology is here to stay."[97] Data would continue to be produced apace, and bioinformatics would continue to be in demand.

> A shift towards an information-oriented systems view of biology, which grasps both mathematically and biologically the many elements of a system, and the relationships among them that allows the construction of an organism, is underway. But the social change required to make this shift painlessly should not be underestimated.[98]

Indeed, the publication of the human genome in 2001 escalated the sense of crisis among biologists to the extent that some feared that traditional biological training would soon become redundant.

> If biologists do not adapt to the computational tools needed to exploit huge data sets . . . they could find themselves floundering in the wake of advances in genomics. . . . Those who learn to conduct high-throughput genomic analyses, and who can master the computational tools needed to exploit biological databases, will have an enormous competitive advantage. . . . Many biologists risk being "disenfranchised."[99]

Again, the "wave" of data created a sense that the old tools were just not up to the job and that a radical re-skilling of biology was necessary. A sense of desperation began to grip some biologists: David Roos no doubt spoke for many biologists when, in the issue of *Science* announcing the human genome draft sequence, he fretted, "We are swimming in a rapidly rising sea of data . . . how do we keep from drowning?"[100] "Don't worry if you feel like an idiot," Ewan Birney (one of the new breed of computationally savvy biologists) consoled his colleagues, "because everyone does when they first start."[101]

This discomfort suggests that bioinformatics marked a radical break with previous forms of biological practice. It was not merely that the HGP used computers to "scale up" the old biology. Rather, what allowed bioinformatics to coalesce as a discipline was that the production and management of data demanded a tool that imported new methods of doing and thinking into biological work. In other words, the com-

puter brought with it epistemic and institutional reorganizations that
became known as bioinformatics. The sorts of problems and methods
that came to the fore were those that had been associated with computing since the 1950s: statistics, simulation, and data management.

As computing and bioinformatics grew in importance, physicists,
mathematicians, and computer scientists saw opportunities for deploying their own skills in biology. Physicists, in particular, perceived how
biological knowledge could be reshaped by methods from physics:

> The essence of physics is to simplify, whereas molecular biol-
> ogy strives to tease out the smallest details. . . . The two cul-
> tures might have continued to drift apart, were it not for the
> revolution in genomics. But thanks to a proliferation of high-
> throughput techniques, molecular biologists now find them-
> selves wading through more DNA sequences and profiles of
> gene expression and protein production than they know what
> to do with. . . . Physicists believe that they can help, bringing a
> strong background in theory and the modeling of complexity to
> nudge the study of molecules and cells in a fresh direction.[102]

Where biology suddenly had to deal with large amounts of data, physicists saw their opportunity. Physicists, mathematicians, and computer
scientists found myriad opportunities in biology because they had the
skills in statistics and data management that bioinformatics required.

Bioinformatics and the HGP entailed each other. Each drove the
other by enabling the production and synthesis of more and more data:
the production of bioinformatic tools to store and manage data allowed
more data to be produced more rapidly, driving bioinformatics to produce bigger and better tools. The HGP was Big Science and computers were tools appropriate for such a job—their design was ideal for
the data management problems presented by the growth of sequences.
Computers were already understood as suitable for solving the kinds of
problems the genome presented. Within this context of intensive data
production, bioinformatics became a special set of techniques and skills
for doing biology. Between the early 1980s and the early 2000s, the
management of biological data emerged as a distinct set of problems
with a distinct set of solutions.

Computers became plausible tools for doing biology because they
changed the questions that biologists were asking. They brought with
them new forms of knowledge production, many of them associated
with physics, that were explicitly suited to reducing and managing large

data sets and large volumes of information. The use of computers as tools of data reduction carried Big Science into biology—the machines themselves entailed ways of working and knowing that were radically unfamiliar to biologists. The institutionalization of bioinformatics was a response to the immediate data problems posed by the HGP, but the techniques used for solving these problems had a heritage independent of the HGP and would continue to influence biological work beyond it.

Conclusions

The sorts of data analysis and data management problems for which the computer had been designed left their mark on the machine. Understanding the role of computers in biology necessitates understanding their history as data processing machines. From the 1950s onward, some biologists used computers for collecting, storing, and analyzing data. As first minicomputers and later personal computers became widely available, all sorts of biologists made increasing use of these devices to assist with their data work. But these uses can mostly be characterized in terms of a speeding up or scaling up: using computers, more data could be gathered, they could be stored more easily, and they could be analyzed more efficiently. Computers succeeded in biology when applied to these data-driven tasks. But biologists often dealt with small amounts of data—or data that were not easily computerizable—so the effects of the computer were limited.

The emergence of DNA, RNA, and protein sequences in molecular biology provided the opportunity for the computer to make more fundamental transformations of biological work. Now biological objects (sequences) could be managed as data. The application of computers to the management and analysis of these objects did not entail merely the introduction of a new tool and its integration into traditional biological practice. Rather, it involved a reorientation of institutions and practices, a restructuring of the ways in which biological knowledge is made. Specifically, the computer brought with it ways of investigating and understanding the world that were deeply embedded in its origins. Using computers meant using them for analyzing and managing large sets of data, for statistics, and for simulation.

Earlier attempts to introduce the computer into biology largely failed because they attempted to shape the computer to biological problems. When the computer began to be used to solve the kinds of problems for which it had been originally designed, it met with greater success. In so doing, however, it introduced new modes of working and redefined the kinds of problems that were considered relevant to biology. In other words, biology adapted itself to the computer, not the computer to biology.

Ultimately, biologists came to use computers because they came to trust these new ways of knowing and doing. This reorientation was never obvious: in the 1970s, it was not clear how computers could be successfully applied to molecular biology, and in the 1980s, Ostell and a handful of others struggled to earn legitimacy for computational techniques. Only in the 1990s did bioinformatics begin to crystallize around a distinct set of knowledge and practices. The gathering of sequence data—especially in the HGP—had much to do with this shift. Sequences were highly "computable" objects—their one-dimensionality and their pattern of symbols meant that they were susceptible to storage, management, and analysis with the sorts of statistical and numerical tools that computers provided. The computability of sequences made the HGP thinkable and possible; the proliferation of sequence data that emerged from the project necessitated a computerization that solidified the new bioinformatic ways of doing and knowing in biology.

I am not arguing here for a kind of technological determinism—it was not the computers themselves that brought new practices and modes of working into biology. Rather, it was individuals like Goad, who came from physics into biology, who imported with them the specific ways of using their computational tools. The computer engendered specific patterns of use and ways of generating knowledge that were brought into biology from physics via the computer. Chapter 2 will characterize these new forms of knowledge making in contemporary biology in more detail, showing how they have led to radically different modes of scientific inquiry.

2 Making Knowledge

How can we characterize bioinformatics? What makes it distinct from other kinds of biology? In chapter 1, I described how computers emerged as a crucial tool for managing data in the 1980s and 1990s. This chapter follows that historical trajectory forward to the present, describing bioinformatic work in action. When I spoke to biologists in seminars and labs, I posed the question, "What is bioinformatics?" The answers were both uncertain and widely varied. Some saw it as a description for the future of all biology, other as an outdated term for a quietly dying field. Some understood it as a limited set of tools for genome-centric biology, others as anything that had to do with biology and computers. Some told me that it was of marginal importance to "real" biology, others that it was crucial to all further progress in the field. Some saw it as asking fundamentally new sorts of questions, others saw it as the same old biology dressed up in computer language. Such confusion immediately piques curiosity: Why should a field evoke such contradictory responses? What could be at stake in the definition of bioinformatics?

These contestations hide deeper controversies about the institutional and professional forms that biology will take in the coming decades. Who gets money, who gets credit, who gets promoted, how will students be trained—

these are high-stakes issues for individual biologists as well as for the future of biology as a whole. Just as mid-twentieth-century physics struggled over its transformation to Big Science, biologists are struggling to sort out their new institutional and epistemic roles. We have already seen how the computational tools and methods described in chapter 1 posed challenges to older forms of biological work and knowledge. Disagreements over tools, methods, and knowledge are still widely apparent. Labels like "bioinformatics," "computational biology," and "systems biology" are important because they denote specific configurations and hierarchies of knowledge, credit, and power within the biology community.[1]

The evidence I gathered from fieldwork and interviews suggests that bioinformatics entails a reorientation of biological practices away from "wet" biological material toward *data*. This may appear to be an obvious claim—*of course* computers deal with data. But what exactly are data for the bioinformatician? Data are defined by the practices of using the computer: they are the stuff that computers manipulate and operate on. They cannot be understood apart from the computer and have no meaning outside of a digital context. The masses of raw, complicated, undigested, disordered stuff that are produced by instruments and experiments become data as they enter and flow through computers. Without computers, they would have no use or significance.[2] So using computers is by definition a data-driven practice.

However, the consequences of this reorientation toward data are more complicated. The data corresponding to an object never constitute a perfect copy of that object. For instance, the data representing a genome are not the same as the "wet" molecular DNA that you find inside your body. But at the same time, it is impossible to conceive of a genome outside a computational context. The work of sequencing and assembling it would have been (and remains) unimaginable without a computer. Even if this somehow could have been achieved, the effort would have been useless without the ability to share and access the data that computers provide. Even the molecular DNA from your body is conceived in terms of its computational or data representation.

To say that biology is becoming about data, then, is to say that it is tied to all sorts of practices, structures, and constraints associated with hardware and software. This claim invites us to investigate how those data are made and used. It suggests that bioinformatics is a distinct and competing set of practices and knowledge for biological work. These practices—derived from computing practices in physics—reorient bioinformatics toward particular sets of questions that are distinct from those of pre-informatic biology. Most importantly, these practices allow biologists to pose and answer *general* questions. Lab bench experimentation usually permits biologists to study only a single gene, a single cell, or a single protein at once. The availability, scope, and shape of data enable questions focused on whole genomes, or hundreds of proteins, or many organisms at once. This is accomplished through the use of statistical methods and approaches: large amounts of data are analyzed and understood *statistically*. These two elements—general questions and statistical approaches—mark a major break with non-bioinformatic or pre-informatic biology. Some of biologists' concerns about this transformation to bioinformatic biology are manifest in debates about "hypothesis-free" or "data-driven" biology.

Divisions of Labor

Where does bioinformatics fit within the field of biology? There is a diversity of ways in which the management of data figures in biological work. Figure 2.1 shows one practitioner's representation of the disciplinary topography. "Biology" includes both computational biology and "systems biology," which overlap with each other. These fields are largely distinct from "informatics," which is broken down into informatics proper ("methods/tools for management and analysis of information/data") and "bioinformatics" ("development of methods for analysis of biological data"). Both of these fields intersect with computational biology, defined as "computational/modeling/analytical approaches to address biological questions." Cutting across all these subdisciplines are the fields of "synthetic biology" and "biological engineering."

This figure represents a particular perspective on a messy and complicated landscape; other biologists would no doubt have their own versions. Bioinformatics is multiple and heterogeneous—not a single entity, but a multitude of different practices. Characterizing it in full requires attending to different scenes and spaces where bioinformatics is taking place. What are those places? First, bioinformatics occupies traditional academic spaces, usually in biology departments, but sometimes distributed between computer science, engineering, chemistry, mathematics, statistics, and medical faculties. Second, semiprivate research institutes such as the Broad Institute and the J. Craig Venter Institute support their own bioinformatic work. Third, government-funded laboratories such as the National Center for Biotechnology Information (NCBI) or the European Bioinformatics Institute (EBI) have bioinformatics as their primary research agendas. Finally, large pharmaceutical companies have

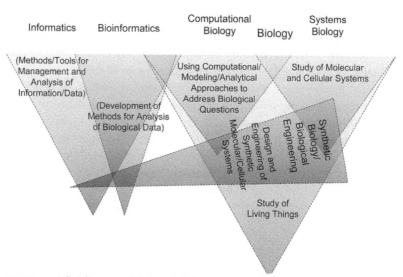

FIGURE 2.1 A disciplinary map of "informatics," "bioinformatics," "computational biology," "biology," and "systems biology" according to one practitioner. (Christopher Burge, "Introduction to Computational and Systems Biology" (7.91), lecture, Massachusetts Institute of Technology, February 8, 2007. Reproduced with permission.)

their own in-house bioinformatics departments supporting their drug discovery work.[3]

If we were to enter these spaces where bioinformatics takes place, what would we observe? To give a sense of the diversity of practice, I describe five scenes that typify work in these various institutional spaces.

∴ ∴ ∴

Scene 1—Office: We enter a building housing an academic biology department. As we climb the stairs to the second floor, we pass by several wet labs where white-coated workers stand at benches with 96-well plates, pipettes, and electrophoresis gels arrayed messily in front of them. At the end of the hallway, however, we find a different sort of space. Here everyone is dressed casually, in jeans and T-shirts; the floor is carpeted, and the lab benches have been replaced with personal computers. The workers sit in front of their monitors, clicking and typing. Indeed, at first glance, the room is indistinguishable from the offices of a software company or Internet start-up: workers are poring over screens filled with numbers (often they are spreadsheets) and text, sometimes examining colorful charts or graphs. This office-like feel of the space is enhanced by a pervasive silence—work is mostly carried out individually. Rarely are the workers looking over one another's shoulders or chatting about the details of their work. Some people enter or leave the room: sometimes a wet-lab worker will come in to ask a question or, more occasionally, to spend some time doing his or her own computer work. Without engaging in the work itself, there is very little else a visitor can observe beyond the flux of text and images on computer screens.

Scene 2—Robo-room: We enter the spacious and plush foyer of an institute devoted to biology. Our host swipes his access card and we ascend to the third floor of the lab. Winding our way amid carpeted offices and shiny-white laboratories, we come at last to a different kind of space: a small doorway leads into a cavernous room with a bare concrete floor. Although there is a mass of equipment scattered about and perhaps as many as ten people in the room, its sheer scale makes it feel almost empty. In the center of the room, fixed to the floor, is a large yellow robot arm, about 1.5 meters high. Around it, arranged in a circle, stand a collection of large pieces of equipment: they are, our host informs us, freezers, incubators, and various other apparatus for adding reagents, mixing, stirring, heating, and growing biological samples. The attention of the workers is devoted to this mass of equipment—some are busy with repairs, bolting or unbolting various pieces to the floor; others are working at laptops and terminal screens connected via fat cables to the robot arm and the equipment circumscribing it. When their work is complete, the humans will be absent from the room altogether: the robot arm will be trained to extract plates of biological samples from the freezers and perform high-throughput experiments by moving these samples between the various pieces of equipment in a carefully prescribed order. The results of the experiments will be electronically communicated to networked hard drives elsewhere in the building where, in turn, they will be accessible to other biologists for analysis. For now, however, the task is to set up the robot and train it to perform its repetitive tasks smoothly and flawlessly.

Scene 3—Production line: This time the building, tucked away in an industrial zone, looks more like a medium-sized factory that a biology institute or academic department. We enter through a service door—there are no identifying signs on the outside of the building and no foyer, just a series of corridors and doors marked with biosafety level warning signs, "BL1," "BL2," or "BL2+." The floors are rubber tiled, and white-coated lab workers push around large carts containing empty beakers or unidentifiable shrink-wrapped supplies. First we encounter a carefully ordered lab space where supplies are arranged neatly on racks against the wall, the space reserved for various machines is marked out with colored tape on the floor, and workers move about the room in a prescribed and repetitive pattern. Observing more closely, we see that the objects of attention are small rectangular plastic "plates." One worker extracts them from a machine and places them in a neat stack on a nearby bench; another worker retrieves them and uses a pipette to add a reagent to each, placing the completed plates in another stack; after some time, a third worker attends to this stack and places the plates on a robotic machine; the plates emerge from the robot in yet another stack and are whisked away by another worker into another room. We follow into a large space filled with dozens of identical machines about the size of a refrigerator. It is almost devoid of people, but a few workers periodically check the computer monitors attached to the machines, making notes on clipboards they carry with them. At certain times, when a machine has finished its "run," workers extract the plastic plates from the machine and carry them away; soon another worker will emerge to replenish the machine with new plates.

Scene 4—Board room: In this scene, we travel to Washington, DC, to the National Institutes of Health campus in Bethesda, Maryland. We proceed into a drab concrete building and ascend to the eleventh floor. Here, in a cramped conference room, about twenty people are gathered, all dressed in business suits. They include the director and grants officers from the NIH, heads of biology institutes, and senior academic biologists. One of the academic scientists begins the meeting with a PowerPoint presentation giving details of his research on the biology of the sea slug. This presentation is followed by another on the recent discoveries concerning human intestinal bacteria. The aim of the meeting is to decide which plant or animal genomes to sequence next and to allocate the money to get it done. The academic biologists are here to advocate for their particular organisms of interest. The representatives of the biology institutes make presentations demonstrating the speed and cost-effectiveness of their past sequencing efforts—they want to secure NIH grant money for genome sequencing. After the presentations, arguments revolve around how best to serve biology by selecting species that will provide appropriate comparisons with known genomes. The discussion comes to a close at 5:30 p.m., and the academic and institute scientists return to their homes; it is now up to the NIH employees to make the final decisions about what to sequence and how to allocate their funds.

Scene 5—Back to the lab: Finally, we enter a space that looks more like a biology lab—pipettes, test tubes, bottles of reagents, and discarded rubber gloves are scattered around the bench spaces. One side of the room is dominated by a particularly large machine: it is a Solexa sequencing machine, a large rectangular box with a transparent panel at the front. The lab workers spend several days preparing DNA samples for this machine, making sure the samples are buffered with the right concoction of chemicals for the sequencing reactions to take place. The samples are inserted into the panel on the front. A "run" of the machine takes several days, so workers go about other work, often preparing more samples for the next run. The results— the data, the sequences—are automatically sent to a computer connected to the sequencing machine. One of the workers now begins to analyze the data. The sequences from the machine must be rigorously compared with sequences contained in online databases. The worker performs a search to compare the sequences with those in GenBank; if the sequences return positive hits against human DNA, then it is likely that the samples were contaminated by DNA from lab workers (since in this case, the sequences are known to come from mouse cells). Next, a more sophisticated set of searches is performed against a range of other databases in order to identify the particular genes and exact genomic loci from which the samples have come.

This diversity of practices makes clear that recent transformations of biology are not merely from noncomputer to computer work. Big Biology now requires large amounts of funding, big laboratories, interdisciplinary teams (computer scientists, mathematicians, physicists, chemists, technicians, and so on), and the expertise to manage all of these. Biologists are increasingly called on to justify their skills and expertise within a diverse landscape of practices and ways of knowing. In particular, biologists have needed to face the challenge mounted by computer scientists who have turned their attention to the analysis of biological problems. This challenge is not just an issue of communication or language, but a struggle over whose modes of doing and knowing will persist in biology.

The stakes in this battle are high: jobs, credit, and money are on line. Will, a veteran of the HGP, told me that when the Whitehead Institute began to scale up for human genome sequencing, it stopped hiring people with degrees in biology and started hiring people with degrees in engineering and experience with management.[4] The director of the Broad Institute, Eric Lander, is a former business school professor with a PhD in mathematics; he has received many accolades for his work on the HGP and is seen as one of the visionaries of the field. Biologists are just getting used to publishing papers in which their names might appear alongside those of a hundred other individuals, including people without degrees in biology.[5] The ideal-typical biologist of the old school—making a contribution to knowledge through toil in his laboratory with only his test tubes and his bacteria to keep him company—is under threat. The use of the term "bioinformatics" by some biologists as a pejorative reflects a discomfort with new forms of biological practice that are massively collaborative: the fact that computers are used is secondary to the fact that they are used to facilitate the collection, distribution, and sharing of information among a wider and wider group of non–biologically trained individuals.

One way in which biologists defend traditional forms of practice is to question the extent to which computational methods can gain access to nature. Computation is all well and good, they argue, but it requires data from "real" experiments, and any conclusions it reaches must ultimately stand up to tests in the wet lab. Without denying the potential power of computational methods for biology, these biologists maintain a hierarchy in which the wet lab becomes the definitive source of authority. The attitude of Charles DeLisi is typical: "You have to prove it in the lab, whatever comes out. . . . So, a computer is useful to generate a kind of hypothesis, . . . biology is, like the physical sciences, strictly empirical, we want observational evidence for proof."[6] Others are quick to criticize work that lacks a sensitivity to wet biology:

> The problem is when the "bio" is missing from informatics. An awful lot of people come from computer science into this field

and all they're interested in is computation and they have no interest in the underlying biology. And so they're busy doing interesting computational things that are biologically completely meaningless.[7]

This comment, made by the Nobel laureate Richard J. Roberts, suggests that the hostility toward computational biology is rooted in a sociological division between individuals with different approaches to biological problems. By name and definition, bioinformatics is a hybrid discipline, built from biology and informatics. Indeed, there are two quite distinct groups of people who inhabit the field. The first consists of individuals with training in the life sciences who have turned to computation in order to deal with larger amounts of data, find more powerful ways to analyze their data, or build computational models of the biological systems with which they work. The second consists of individuals who have been trained in computer science, mathematics, or engineering (or in computational areas of physics) and who have turned to biology to find new kinds of problems to solve and new applications for their programming and algorithm-building skills.[8] As one might predict from Roberts's comments, these groups have different interests, different approaches to problems, and different attitudes about what constitutes good biological work. They may occupy different academic departments and publish in different journals.[9]

In fact, almost everyone I talked to in the fields of bioinformatics and computational biology complained about problems of communication across the computer science–biology divide. When I asked what challenges awaited bioinformatics in the next few years, hardly anyone failed to mention the fact that more people needed to be adequately trained in both biology and computer science. Much of what was characterized as "bad" or "useless" work in the field came either from computer scientists misunderstanding what problems were important to biologists or from biologists failing to realize that the computational problems with which they were engaged had been solved by computer scientists decades ago.

Christopher Burge, in whose lab I worked, was unusually respected in the field because his background in mathematics and computer science was complemented by an encyclopedic knowledge of biology; as such, he was able to run both a computational lab and a wet lab side by side. But even so, Burge had to pay constant attention to the disjunctions within the field: he had to ensure that his lab was balanced between

PhD students and postdocs with computational and wet-lab skills, and he had to worry about how frequently his lab members moved back and forth between the two domains—too much or too little time on one side or the other might damage their chances for doing successful biological work.[10] And, most importantly, he had to ensure that members of the wet lab and computer lab were communicating well enough to collaborate effectively. Doing good work meant continually attending to this boundary between computation and biology.

During my fieldwork, a PhD student from the computer science department at MIT visited Burge's lab. As he explained to me in an interview, he was trying to find a "good" problem for his dissertation work, so his advisor had made arrangements to have him spend some time in a biology lab, looking for an appropriate biological problem to which he could apply his computational skills. At least in the beginning, he had found such a problem elusive: "It's very hard to get a PhD level background in computer science and biology. Right now I'm doing this project in a biology lab but . . . I don't feel like I have a huge amount of background, so at this point I'm trying to figure out the things that I need to know to get the work done." Part of the reason for the problem, he surmised, was a fundamental difference in the way biologists and computer scientists think:

> Most computer scientists are probably . . . put off of biology because the very things they like about computer science seem to not be present in biology . . . Francis Collins even made that particular observation, when he first started his career he was a physicist, I believe, and he stayed away from biology because it was quote-unquote messy—it was messy science and he didn't want to deal with the messy details. . . . A lot of computer scientists who may want to get into computational biology would have to go through a similar process of reconciling their dislike for messiness, because they are trained to look for elegance, mathematical elegance.

This became a constant trope in my interviews and conversations with biologists and computer scientists: the latter were interested in elegant, discrete, neat, tractable problems, while the former constantly had to deal with contingency, exceptions, messiness, and disorder.

The lack of communication between computer scientists and biologists, then, goes deeper than problems with language. Indeed, we can understand the rift between them as stemming from commitments to fundamentally different sorts of practices. In order to explore these differences, let me give a more detailed example of each:

: : :

The biologist: Natalie is interested in the problem of alternative splicing. In particular, she wants to determine how many alternative splicing events are conserved between human and mouse, since the result would suggest the importance of such events for biological fitness. First, she downloads the genomes of both human and mouse from GenBank. Then she downloads large collections of EST (expressed sequence tag) data from publicly available databases for both human and mouse. Natalie uses a freely available piece of software called GENOA to align the ESTs with the human and mouse genomes. This results in almost a million human EST alignments and about half a million mouse EST alignments. Natalie then writes a program to determine which of these ESTs, for both mouse and human, contain alternative splicing events. She uses online databases (EnsMart and Ensembl) to determine which pairs of human and mouse genes match one another (are orthologous). This procedure allows Natalie to identify about two thousand alternatively spliced EST events that were conserved between human and mouse. Wishing to validate these results in the wet lab, Natalie then performs experiments using RT-PCR (real-time polymerase chain reaction) on RNA from human and mouse tissues, verifying about thirty of the alternative splicing events.[11]

The computer scientist: Henry is also interested in alternative splicing. However, he conceives the problem as one of predicting de novo which sequences, within a very large sample, map to exon junctions in the human genome. First, he computes the parameters of the problem, demonstrating that a brute force search would require 72 million CPU-years. This finding provides the justification for developing a novel algorithm to speed up the process. After developing an algorithm he thinks will achieve a significant speedup, Henry spends several weeks "prototyping" it in Java. Although the algorithm is successful, it does not run as fast as Henry requires. He decides to reimplement the algorithm in the C language, which allows him much finer control over memory access. Henry also spends a further two weeks "profiling" his code: he runs a program on his code while it is running to analyze where it is spending the most time—which lines are acting as bottlenecks. Once bottlenecks are identified, Henry works to speed up these parts of the code, perhaps increasing its speed by a factor of 2 or 3 here and there. After about eight weeks, with the revised algorithm now running in C, Henry predicts that his code will run on the data set he has in mind in about two or three CPU-months. On a powerful computer cluster, this amounts to mere days. The problem is now tractable, and Henry's work is done.[12]

Both of these sets of activities are primarily computational, yet the biology-trained student's approach differs markedly from the computer science–trained student's approach. For the former, the computer programs are not ends in themselves—it is sufficient that they work at all. Biology-trained individuals are often "hackers," piecing together results by using whatever resources they have at hand (other people's code, the Internet). The computer scientist, on the other hand, considers programming an art: the final product (the code) should provide an elegant solution to the problem at hand. This difference is manifest in the different sorts of practices we see in this example: the biology trainee spends the most time writing small chunks of code that help to order and make sense of the data; the computer scientist is interested in the whole product and spends the most time revising it over and over again to find an optimal solution to the problem. The status of wet biology is also understood very differently: for the biologically trained, the wet lab must be the ultimate arbiter of any finding, whereas the computer scientist, while acknowledging its importance, produces results that are designed to stand on their own.

These differences in approach can be understood as two distinct attitudes toward data: for the computer scientist, the point of his or her work is the elegant manipulation of data; for the biologist, the wet stuff of biology takes precedence over data. Moreover, the computer scientist's work is oriented toward mastering and digesting the whole quantity of data, which comes down to a problem of managing the quantity of data by increasing speed. The biologist, on the other hand, is more concerned with the particularity of specific cases: it might take years of work to validate every result shown by the data in the wet lab, so more work must be done to narrow down the data to biologically plausible cases (those conserved across species); only a few of these cases were chosen as representative samples to be verified by RT-PCR. The biologist's work is still oriented toward the lab bench and the specificity of particular samples. The computer scientist, on the other hand, is aiming to digest more and more data at higher and higher speeds. This is not a conflict about whether or not to use computers, but rather a battle over different ways of working and knowing in biology.

Mass and Speed

The examples given in the previous section might be placed along a spectrum: more traditional biological work lies at one end, while newer computational approaches lie at the other. One way of placing work along such a spectrum would be to characterize it in terms of *mass and speed*. Whether we consider a high-throughput sequencing machine, an automated laboratory, or a high-level meeting about research grants, bioinformatic biology is obsessed with producing more data, more rapidly. It possesses an insatiable hunger for data. Its activities are oriented toward *quantity*. Data production, processing, analysis, and integration proceed more and more rapidly: the faster data can be integrated into the corpus of biological information, the faster more data can be produced. Data production and data analysis are bound into a feedback loop that drives a spiraling acceleration of both.

In a conference held at the NCBI for the twenty-fifth anniversary of GenBank, Christian Burks (a group leader at Los Alamos when Gen-Bank was started) asked, "How much sequencing is enough?" Speaking only slightly tongue in cheek, Burks began to calculate how many base pairs existed in the universe and how long, using current technology, it would take to sequence them. Although this, he admitted, represented a ridiculous extreme, Burks then computed how long it would take to sequence one representative of every living species, and then predicted that we would eventually want about one hundred times this amount of sequence data—the equivalent of one hundred individuals of each living species.[13] No one in the audience seemed particularly surprised. The ability and the need to sequence more, faster, had become obvious to at least this community of biologists.

Another example of this voracious need for data is provided by the work of Graham. Graham's work uses genomic and other data to try to understand how and why errors occur in protein folding. Closely related species usually possess closely related proteins, perhaps with only a few nucleotide differences between them. Yet, since both species exist, both versions of the protein must fold correctly, or at least function in the same way. By examining many such closely related proteins, Graham can build up a statistical theory of what kinds of changes can be made to a protein without causing it to fail to fold. However, each pair of proteins can contribute only a tiny amount of information about possible nucleotide substitutions, and therefore many protein pairs are required for this kind of analysis. As such, much of Graham's work is in constant need of more sequence data:

> The strategy I'm trying to adopt is: let's take some of the classic tests and measures [for the adaptation and evolution of proteins] and recast them in ways where, if we had all the data in the world, then the result would be clearly interpretable. The *only* problem that we have is not having all the data in the world.

Graham's work proceeds as if "not having all the data in the world" is a problem that will soon be solved for all practical purposes. Soon after the interview quoted above, the Broad Institute published the genomes of twelve different species of fruit flies, providing a gold mine of data for Graham's work.[14]

The growth of sequence data has become so rapid that biologists commonly compare it to Moore's law, which describes the exponential growth of computing power (in 1965, Moore predicted a doubling every two years). In 2012, the quantity of sequence was doubling roughly every six months. Some argue that, in the long run, sequence data may outstrip the ability of information technology to effectively store, process, or analyze them.[15] However, this argument overlooks the fact that growth in computing power (processing power and storage) may drive the growth in biological data—it is at least in part because of Moore's law that biologists can imagine "having all the data in the world" as a realizable goal. There is a feedback loop between Moore's law and biological work that drives biology toward an ever-greater hunger for data and the ability to process them instantly. Over the duration of my fieldwork, the greatest excitement in the field (and probably in all of biology) came not from a new discovery in evolution or genetics or physiology, but rather from a new set of technologies. The buzz about the so-called next-generation (or next-gen) sequencing technologies dominated the headlines of journals and the chatter in the lunchroom.[16] In countless presentations, the audience was shown charts illustrating the growth in sequence data over time or (conversely) the reduction in cost per base pair produced (figure 2.2). Biologists clamored for these new

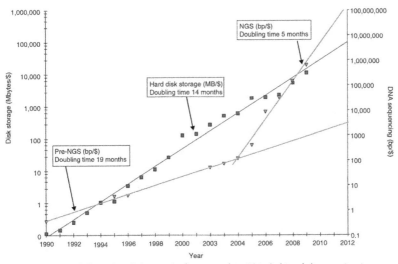

FIGURE 2.2 Typical illustration of the growth of sequence data. Historical trends in computer storage prices are plotted against DNA sequencing costs. This is a logarithmic plot, meaning that the straight lines correspond to exponential growth. The cost of computer storage (squares) is shown as growing exponentially with a doubling time of roughly 1.5 years (equivalent to Moore's law). The cost of DNA sequencing (triangles) is shown in two separate lines corresponding to pre- and post-NGS (next-generation sequencing). For the latter, the doubling time reduces to six months. (Stein, "The case for cloud computing." Reproduced from *Genome Biology* under the BioMed Central open access license agreement.)

machines, heckling their directors and department chairs to authorize the expenditure (about $350,000 for a single machine plus about $10,000 a week for the chemicals to keep it running continuously) lest they be unable to continue to do biology at the cutting edge.[17]

But this need for quantity extended from the technology into the work itself. During my visit, the Burge lab was devoted to learning about the nature of RNA splicing. In particular, the group wished to discover how RNA splicing was regulated and what role this regulation played in protein function.[18] What was interesting about the Burge lab's work, however, was that it hardly concerned itself at all with particular splicing events. Instead, the questions the team wished to answer concerned the nature of *all* splicing events. How many genes in the genome are spliced in a particular way? How often? How many ways can they be spliced on average? Are there any patterns in which exons get skipped? What can we learn about the ways splicing is regulated from the *totality* of events?[19] Answers to such questions require an immense amount of data—the Burge lab needed multiple sequences of the same gene extracted from different tissues at different times, under a variety of conditions. Only next-gen machines could make such data available.

The consequence of this obsession with speed and quantity is that biologists are asking different kinds of questions. In the Burge lab, the specific nucleotide, the specific gene, the specific genome, even the specific
disease, move into the background. It is the *general* problems that are
at issue: How do all genes work? What are the overall rules for how genomes or diseases behave? Many informants remarked to me that they
perceived this as a major break with other kinds of biology that deal
with single proteins, single genes, and single species, which require years
of work at the bench to determine the particularities and peculiarities of
a locally defined object. I asked one graduate student in biology why he
found his work on gene regulation interesting. He replied,

> I think that it's an interesting problem because you can look
> at global changes; you can look at every gene in a cell at once.
> Whereas, with a lot of other kinds of biology you have to break
> things down into manageable problems, you can't look at things
> on a whole system level, you can't look at the biological pro-
> cesses all together because it's just not possible with the tools
> that we have. . . . So, because of having all these tools and all this
> data, you're able to look at the big picture.

But he was also quick to admit that the questions were also driven by
the availability of large data sets and the computational power to analyze them:

> **Student:** It's kinda circular: part of the reason why I think it's
> interesting is because it's a good problem for me to solve . . .
> it's becoming feasible to answer the question . . .
> **HS:** Because there's so much data?
> **Student:** Yeah, because of the methods that are available . . .
> full genome sequencing, microarrays, the new faster sequencing
> methods . . .

The kind of biology I am describing here, then, is driven by quantity;
it relies and depends on a constantly increasing stream of data. Data
drive biology toward general, rather than local and specific, problems,
and for that reason some have identified this kind of biology as "theoretical."[20] However, this reorientation is less a result of a shift toward
theory than a result of the culture of quantity that has precipitated new
kinds of questions. It was not directly *caused* by sequencing or by microarray or other technologies. Rather, the technology and the culture are

engaged in a dialectic in which each drives the other: the valorization of mass, volume, and speed lead to a greater demand for data, which justifies better and faster sequencing, which, in turn, contributes to the continued success of the "quantitative" approaches.

Michael Fortun has drawn attention to the cultures of "acceleration" and "speed" that surrounded the Human Genome Project. He argues that much of the controversy surrounding the project was connected to its speed—many wondered why the HGP was considered to be so urgent. The commitment to the HGP involved a commitment to prioritizing (and therefore speeding up) certain kinds of biological work in particular labs.[21] Bioinformatics has become the space in which the conflict between these older and newer (slower and faster) forms of practice ultimately plays out. Already by the mid-1980s, "informatics was becoming the most important speed nexus";[22] computers were crucial for the HGP not only to achieve automation, but also to manage and maintain the speed of data production.[23] The continued drive toward rapid and large-scale biology was created not by the HGP, but by computers.

More data create not just more knowledge, but a qualitatively different kind of knowledge. Consider, for example, the type of scientific publication routinely produced by the Broad Institute. In 2005, the lab published a paper based on a draft of the chimpanzee (*Pan troglodytes*) genome. The sequencing work alone—performed at both the Broad and the Washington University School of Medicine—was not considered publishable material in and of itself. Indeed, the chimpanzee sequence had been uploaded to public databases as it was completed and had been available for some time. Rather, the "chimpanzee genome paper" consisted largely of analysis and comparative work performed *on* the sequence itself. The work of the Broad team consisted of building software, aligning the sequence, and statistically analyzing the data. The result was a high-level overview of the features of the chimpanzee sequence. This published knowledge is not raw data, nor is it anything like detailed analysis of a particular gene or cell or protein; rather, it is an attempt to fully characterize the genome of an organism at a high level.

The Broad's techniques aim to understand the genome in its totality—to draw conclusions about all genes, or all transposons, or all repeating elements in a genome. Software and statistics generate a high throughput of knowledge, an efficient transformation of data into publications. Those at the Broad who were involved in this work related how they thought that computers had changed the questions that were asked in this sort of biological work—the emphasis falls on general problems, on big patterns and correlations. The work of the Broad team becomes to analyze and make sense of a vast output of data: comparing genomes, describing the whole human microbiome, analyzing and classifying all forms of cancer, mapping the totality of human haplotypes, and so on.[24] One informant vividly described this kind of task as creating a "Google Earth for looking at the data."

Whereas pre-informatic biologists usually dealt with a single organism, or perhaps even a single gene—devoting their careers and the careers of their graduate students to characterizing it in detail—bioinformatic biologists ask (and answer) bigger questions—questions that rely on knowledge of all human genes, or a comparison across twenty or more organisms. In short, they depend on having extraordinarily large volumes of reliable and accessible data. Indeed, the biologists at the Broad are in constant need of more sequence data. Although it is true that more traditional biologists are also interested in general questions about life, their day-to-day focus usually remains specific; without genome-scale data (and bioinformatic techniques to digest them), answers to such questions must rely on speculative extrapolation from a few examples. The big questions remain the same, but the answers that can be given are of a markedly different kind.

If molecular biology was about attention to detail and particularity, bioinformatics is about generality and totality. Whereas a molecular biologist might spend a career working to characterize a particular gene, asking questions about its function, its regulation, and its interactions, the bioinformatician is interested in *all* genes, or *all* proteins, or *all* interactions. Using computers means that it is possible to ask new, broader kinds of questions. But working at this level requires a surrendering of certain kinds of particular, granular knowledge. Using bioinformatics to determine that 45% of genes in humans are regulated by two or more different regulatory proteins tells the investigator nothing about which particular proteins regulate a particular gene. Bioinformatics entails a decision that certain types of knowledge are not important—that a certain amount of ignorance of detail is required. Biologists' response to a growing amount of biological data was to find a different route to turn information into knowledge—there were too many data to continue to use and process them in the way that molecular biology had done since the 1950s. Instead, biologists began to use computers to manage what could not be known about the details of particular genes, proteins, and so on.

Paying attention to what bioinformatics does with data shows how the computer has imported new techniques into biology. These techniques have, in turn, engendered new approaches directed toward answering *general* or *high-level* questions about biological systems.

Bioinformatics at Work

But this discussion of the scope of bioinformatics fails to provide any account of *how* bioinformatics does its work. How are the data *actually used* to answer these big questions? What do computers actually do to reduce vast amounts of data to meaningful knowledge? Answering these questions requires a closer look at what is going on inside a computational biology lab. Following the work of the Burge lab in detail demonstrates the importance of statistical approaches in this work. The lab used computers not only to analyze data, but also in specific ways as tools for statistics and simulation.

During my fieldwork, the Burge lab began collaborating with Illumina Inc., the manufacturer of the brand-new (in 2007) next-generation Solexa high-throughput sequencing machines. Illumina hoped to demonstrate the usefulness of its machines by providing data to academic biologists who would analyze those data and publish their findings. The Burge lab was happy to accept a large volume of new data. The Solexa machines were able to produce an unprecedented amount of sequence data in a remarkably short time. At a practical level, the collaboration with Illumina had provided the Burge lab with a data set of such size and richness that the team felt they did not have the resources to analyze it completely. This data glut presented a fortunate circumstance for my fieldwork: plenty of interesting projects and problems were available on which I could cut my teeth.

The data themselves were gene expression data from four distinct tissues in the human body (heart, liver, brain, skeletal muscle). That is, they consisted of millions of short sequences of mRNA. DNA is transcribed into mRNA (messenger RNA) before being translated into proteins, so a collection of mRNA sequences provides a snapshot of the genes that are being expressed (that is, made into proteins) in particular cells at a particular time. Before Solexa technology, amassing an equivalent set of expression data would have been a significant experimental program in its own right.[25] Solexa machines produced the data automatically from a few runs of the machine over a few days.

In all higher organisms, the mRNA transcript is spliced: introns are cut out and discarded, and exons are stitched together by the cellular machinery. Since the Solexa sequencing took place post-splicing, the sequences could be used to study the splicing process itself—the Burge lab was particularly interested in how the splicing process is regulated by the cell. In general, splicing is straightforward and linear: exon 1 (the first exon appearing along the linear length of the DNA molecule) is spliced to exon 2, which is spliced to exon 3, and so forth. Sometimes, in so-called alternative splicing events, exons can be skipped or introns can be retained: exon 1 is joined to exon 3, which is joined to exon 4, and so on; or the intron between exon 1 and exon 2 is not excised. Alternative splicing is a well-known, if little-understood, process. More rarely, stranger things seem to happen. Exons can be repeated: exon 1 is joined to exon 1. Or exons seem to be spliced out of order: . . . exon3-exon1-exon2-exon3-exon4 . . . Such events had been observed, albeit rarely, in studies of individual genes.[26] The Solexa data provided one of the first opportunities to study such events—sometimes called exon scrambling—on a genome-wide level, not just as peculiarities of particular genes.[27] During my fieldwork in the Burge lab, I was given the task of using the Solexa data to see what they might reveal about exon scrambling. This was, first, a classic bioinformatic problem in the sense that the aim was to analyze exon splicing in general, on a whole transcriptome, not as it applied to specific genes. Second, it was an ontological problem—an attempt to find out whether exon scrambling was a "real" biological occurrence or whether it was some sort of artifact of sequencing or other laboratory techniques.

The general approach that I was instructed to take was to write small computer programs to find possible scrambling events and then search for any regular patterns in their distribution. The idea was that if scrambling events were showing up as some sort of artifact, then they should be randomly distributed with respect to biological features. If, on the other hand, scrambles occurred more often (than predicted by chance) near long exons or short exons, or near the beginnings or ends of genes, or on especially long or especially short genes, or near particular kinds of splice sites, or on genes with particular functions, that would provide evidence that they were "real." For instance, a comparison of the lengths of introns that stood adjacent to scrambled exons with the lengths of all introns in the genome showed that introns near the scrambling events were significantly longer than average. This finding suggested that some biology was at work.

Although understanding the splicing mechanism was important, using statistics was much more crucial than using biology in these investigations. In particular, I had to go to great lengths to devise a way to show that scrambling events deviated from randomness in a statistically significant way. One way was to create a "decoy" set of exon scrambling events: an imaginary set of scrambling events of the same size as the actual set, but distributed randomly across exons for the given genes. The idea was that if a biological effect showed up in the true set of scrambling events and not in the decoy set, then I was observing a real effect; if it showed up in both, something else, probably nonbiological, was going on.

At the end of my fieldwork, when I presented my findings at the weekly lab meeting, most of the criticisms I received related to statistical shortcomings. Indeed, the suggestions for improving my analysis centered on ways to make better statistical models of scrambled exons in order to differentiate signal from noise. In other words, it was not so much that my work had left room for other plausible biological explanations as that it had not ruled out other possible *statistical* explanations. This example suggests that bioinformatics entails new criteria for evaluating knowledge claims, based on statistical, rather than direct experimental, evidence. But it also shows the importance of the computer as a tool for rearranging objects, rapidly bringing them into new orders and relationships: mixing up, sorting, and comparing chunks of sequence.

This project, as well as other work I observed in the lab, can also be considered a form of simulation: scrambled exons could be considered "real" events only when a model (the decoy set, for instance) failed to produce the observed number of scrambling events. This process is remarkably similar to the accounts of particle physics experiments referred to in chapter 1—there too, a simulation of the background is generated by the computer and the particle is said to be really present if the observed number of events is significantly above that predicted by the simulation.[28] Early in my fieldwork, I jotted down the following observation:

> The work in the lab seems to me to have a lot of the messiness of other lab work. Just as you can't see what proteins or molecules are doing when you are working with them on a lab bench, in this work one cannot see all your data at once, or really fully understand what the computer is doing with it. There are analogous problems of scale—one realm is too small, the other too informatically vast for you to try to manipulate individual pieces. This means that you have to try to engineer ways to get what you want out of the data, to make it present itself in the interesting ways, to find the hidden pattern, just as you

could find the hidden properties of molecules by doing some sort of biochemical assay.[29]

Most bioinformatic problems have this character—there are too many data points, too many individual cases, to check one by one; only the computer has the time and the power to process them. But how do you decide when something is "real"? How do you decide whether the computer is producing a biological result or an artifact? How do you trust the machine?

When I posed this question to bioinformaticians, or when I observed their work in the lab, it seemed that they were using the computer itself to reassure themselves that it was producing "biology" rather than garbage. This process was often described as a "sanity check"—tests were specially designed to check that the computer was not taking the data and doing something that made no sense. Bioinformaticians consider these tests to be the computational equivalent of experimental controls. In a traditional wet lab, a control is usually a well-characterized standard—if you are measuring the length of a sequence on a gel, the control might be a sequence of known length. If the experiment does not reproduce the length of the control accurately, then it is unlikely that it is measuring the lengths of unknown samples accurately either. A computational sanity check could be described in the same terms: run the computer program on a well-characterized data set for which the result is known to check that it is giving sensible results. For example, sometimes it is possible to use the computer to construct an alternative data set (often using random elements) that, when run through a given piece of software, should produce a predictable outcome. Just as with my project, a large part of the work of doing bioinformatics revolves around the construction of appropriate controls or sanity checks. This is not a trivial task—the choice of how to construct such tests determines the faith one can put in the final results. In other words, the quality and thoroughness of the sanity checks are what ultimately give the work its plausibility (or lack thereof).

Such sanity checks are actually simulations. They are (stochastic) models of biological systems, implemented on a computer. Bioinformaticians are aware that there is much about both their data and their tools that they cannot know. As for the data, there are simply too many to be able to examine them all by eye. As for their tools, they necessarily remain in ignorance about the details of a piece of software, either because it was written by someone else or because it is sufficiently complicated that it may not behave as designed or expected. In both cases, it is not simply a question of "opening up a black box" or doing more work to fill in the blanks: starting to explore the data by hand or beginning to analyze a piece of software line by line is usually a practically impossible task.[30] This ignorance has many sources—it may come from the proprietary nature of software or lab instruments, from the sheer size of data sets, or from inherent uncertainties in an experimental technique.[31] In bioinformatics, though, this ignorance is controlled through a specific process of computational modeling: sanity checks or simulations fill in the gaps. It is through these specific computational techniques that ignorance is controlled and trust in computational results is produced.

This kind of knowledge production marks a significant break with pre-informatic biology. The general questions that bioinformatics poses can be addressed by using computers as statistical and simulation tools; the management of large amounts of data enables bioinformaticians not only to ask big questions, but also to use the computer as a specific type of data reduction instrument. Bioinformatic biology is not distinct simply because more data are used. Rather, the amounts of data used require distinct techniques for manipulating, analyzing, and making sense of them. Without these computerized statistical techniques, there would be no data—they would simply not be collected because they would be worthless. These techniques make it possible not only to aggregate and compare data, but to parse, rearrange, and manipulate them in a variety of complex ways that reveal hidden and surprising patterns. Computers are required not only to manage the order of the data, but also to manage their disorder and randomness.

Hypothesis and Discovery

In the last decade, several terms have arisen to label the kind of work I am describing: "hypothesis-free," "data-driven," "discovery," or "exploratory" science. A traditional biologist's account of knowledge making might come close to Karl Popper's notion of falsification:[32] theory firmly in mind, the biologist designs an experiment to test (or falsify) that theory by showing its predictions to be false. However, this hypothetico-deductive method can be contrasted with an older notion of inductive science as described by Francis Bacon. According to Bacon, scientific inquiry should proceed by collecting as many data as possible prior to forming any theories about them. This collection should be performed "without premature reflection or any great subtlety" so as not to prejudice the kinds of facts that might be collected.[33]

Bioinformatics can be understood in these terms as a kind of neo-

Baconian science in which hypothesis-driven research is giving way to hypothesis-free experiments and data collection.[34] Such work involves the collection and analysis ("mining") of large amounts of data in a mode that is not driven by particular hypotheses that link, say, specific genes to specific functions. Genome-wide association studies (GWAS), for instance, search for genes associated with particular phenotypes by searching large amounts of genotypic and phenotypic data for correlations that statistically link specific genomic loci to particular traits. With enough data, no prior assumptions are needed about which loci match which traits. This sort of computer-statistical analysis can generate relationships between biological elements (pieces of sequence, phenotypes) based not on shape, or function, or proximity, but rather on statistics. In other words, the data-statistical approach provides new dimensions along which biological objects can be related.

The examples in this chapter demonstrate that bioinformatics is directed toward collecting large amounts of data that can subsequently be used to ask and answer many questions. Thus the terms "data-driven" and "hypothesis-free" have become focal points of debates about the legitimacy of bioinformatic techniques and methods. Both biologists and philosophers have taken up arguments over whether such methods constitute legitimate ways of producing biological knowledge. The contrasting positions represent tensions between new and old forms of knowledge making and between new and old forms of organizing biological work. Indeed, the sharpness of these epistemological disagreements is further evidence that bioinformatics entails a significant challenge to older ways of investigating and knowing life.

The biochemist John Allen has been especially outspoken in arguing that there can be no such thing as hypothesis-free biology and even that such work is a waste of time and money: "I predict that induction and data-mining, uninformed by ideas, can themselves produce neither knowledge nor understanding."[35] Alarmed that hypothesis-driven science is portrayed as "small-scale" and "narrowly focused," critics of the data-driven approach argue that true science is not possible without a hypothesis, that a lack of a hypothesis indicates a lack of clear thinking about a problem, and that the data produced by hypothesis-free science will be uninterpretable and hence useless.[36] Sydney Brenner, another particularly outspoken critic, describes data-driven biology as "low input, high throughput, no output." In his view, both systems biology and genome-centered biology work at the wrong level of *abstraction*— simply examining more data at a genome-wide level will not help biologists understand how organisms work.[37] In other words, what bothers Brenner is the generality of the questions that are being posed and answered.

A significant proportion of the debate over hypothesis-free biology has centered on the computer and its capabilities as an "induction machine." Allen in particular argues that discovery science amounts to letting the computer "do our thinking for us": "knowledge does not arrive *de novo* from computer-assisted analysis . . . we should not give [computers] credit for having the ideas in the first place."[38] Such views suggest that much of what is at the root of the discomfort with discovery science is the role of computers in reshaping biological practice and knowledge production. Indeed, without the ability to process, reduce, analyze, and store data using computers, hypothesis-free science would be impossible. It was the introduction of the computer to biology that enabled the shift from hypothesis-driven or deductive methods to inductive methods within biology.

However, it is not merely the use of computers itself that is important; rather, it is the use of the computer as a specific tool for the management and analysis of large quantities of data using statistical techniques.[39] Biology *in silico* is not just a consequence of large data sets, or of the availability of computers. Rather, it is a consequence of the specific epistemologies, practices, and modalities of knowing that were and are embedded in the transistors and chips of computing machines.[40] The discomfort with bioinformatics and hypothesis-free biology is due to the specific ways of using the computer (general questions and statistics) that I have described in this chapter.

The conflict between hypothesis-free and hypothesis-driven biology has a longer heritage in debates about the "data crisis." As already noted, since at least the beginnings of GenBank, there has been a persistent concern that biology will soon be overwhelmed by data that it will be unable to process or analyze. In 1988, before the HGP had gotten under way in earnest, Charles DeLisi, one of its strongest advocates, expressed this worry:

> The development of methods for generating [sequence] information has, however, far outstripped the development of methods that would aid in its management and speed its assimilation. As a result, we are witnessing enormous growth in data of a most fundamental and important kind in biology, while at the same time we lack the ability to assimilate these data at a rate com-

mensurate with the potential impact they could have on science and society . . . the problem could rapidly grow worse.[41]

Such warnings have been repeated over and over again by biologists who believe that it's time to stop accumulating data and time to start analyzing the data that already exist. Insights into the workings of biology, and potential medical or pharmaceutical breakthroughs that might emerge from this knowledge, are being missed, the argument goes, because they are buried in the flood of data. The data, like an ocean, constitute an undifferentiated mass that cannot be tamed into knowledge. But this dilemma applies only to older forms of biological work in which data were to be analyzed one by one. Bioinformatic modes of work *can* tame vast amounts of data—in fact, for some forms of bioinformatics, the more data the better.

The philosopher of science Laura Franklin-Hall has drawn a useful distinction between experiments using "narrow" instruments, which make a handful of measurements in order to test specific hypotheses, and "wide" instruments, which make thousands or even hundreds of thousands of measurements. Whereas a Northern blot might be used to measure the presence and abundance of a couple of species of mRNA, a DNA microarray is capable of measuring the levels of thousands of mRNAs at once.[42] The microarray's inventors advocate just this type of novel use for their instrument:

> Exploration means looking around, observing, describing, and mapping undiscovered territory, not testing theories or models. The goal is to discover things we neither knew nor expected, and to see relationships and connections among the elements, whether previously suspected or not. It follows that this process is not driven by hypothesis and should be as model-independent as possible . . . the ultimate goal is to convert data into information and then information into knowledge. Knowledge discovery by exploratory data analysis is an approach in which the data "speak for themselves" after a statistical or visualization procedure is performed.[43]

This notion of letting the data "speak for themselves" is no doubt a problematic one: all kinds of models, assumptions, and hypotheses necessarily intervene between a measurement and a certified scientific fact. Yet it accurately portrays what many biologists think they are doing in bioinformatics: setting the data free to tell their own story. Because such data are not understood to be tied to a specific hypothesis, they may have many uses, many voices, many stories to tell. This is a fundamentally different mode of investigation that is increasingly central to biological work.[44]

Franklin-Hall has also suggested that part of what is going on with exploratory science is a maximization of efficiency in scientific inquiry. Where a narrow instrument dictates that only one or two measurements can be made at once, it pays to think very carefully about which measurements to make. In other words, it is efficient to use the instrument to test a very specific hypothesis. With a wide instrument, however, this restriction is removed, and efficiency of discovery may well be maximized by making as many measurements as possible and seeing what interesting results come out the other end. However, this economy of discovery can be realized only if the investigator has some means of coping with (processing, analyzing, storing, sharing), the vast amount of data that are generated by the instrument. This is where the computer becomes the crucial tool: efficiency is a product of bioinformatic statistical and data management techniques. It is the computer that must reduce instrument output to comprehensible and meaningful forms. The epistemological shift associated with data-driven biology is linked to a technological shift associated with the widespread use of computers.

Natural philosophers and biologists have been collecting and accumulating things (objects, specimens, observations, measurements) for a long time. In many cases, these things have been tabulated, organized, and summarized in order to make natural knowledge.[45] But specimens, material objects, and even paper records must be treated very differently than data. Data properly belong to computers—and within computers they obey different rules, inhabit different sorts of structures, are subject to different sorts of constraints, and can enter into different kinds of relationships.[46] Bioinformatics is not just about *more* data, but about the emergence of data into biology in tandem with a specific set of practices of computerized management and statistical analysis. These computer-statistical tools—expectation maximization, Bayesian inference, Monte Carlo methods, dynamic programming, genetic algorithms—were exactly the ideas that began to be taught in the new bioinformatics courses in the 1990s. These were the tools that allowed biological knowledge to be built out of sequence. These were the tools that have allowed biologists to reorient toward general, large-scale questions. Rather than studying individual genes or individual hypotheses one at a time, bioinformaticians have replaced this linear approach with tools that search for patterns in large volumes of data. They are generating knowledge out of the seeming randomness and disorder of genomes.

When computers were brought into biology in the 1980s, they brought with them certain patterns of use, certain practices, particular algorithms, and modes of use that persisted in biological work. It was not that the machines themselves dictated how they were to be used, or the kind of biology that could be practiced with them, but rather that their users had established ways of working and well-known ways of solving problems that could and would be deployed in organizing and analyzing biological data. These ways of working have conflicted with and challenged older forms of doing and knowing in biology.

This conflict has arisen partly because the shift between old and new has not been merely a technological one. Bacon's vision of science extended to a rigidly structured society, working toward the pursuit of knowledge through a strict division of labor. One group of individuals would collect data, conducting a large and diverse range of experiments; a second group would compile the results of these experiments into tables; and a third would construct theories based on these tabulated observations.[47] This chapter has shown that hypothesis-free biology is not merely about a shifting epistemological approach to scientific inquiry—rather, it is entailed by and intertwined with changes in technologies, changes in disciplinary boundaries, changes in institutions, changes in the distribution of funding, and changes in the organization of work and workers. In other words, Bacon was right in suspecting that the production of a particular kind of knowledge required a particular organization of people and spaces.

In chapters 3 and 4 we will see how bioinformatics has driven just these kinds of shifts in the organization of biological work. The emergence of bioinformatics meant the emergence of data in biology. Data cannot be understood separately from computers and the ways in which they are used: they entail the asking of specific sorts of questions and enable particular sorts of solutions. Data are not like a specimen or a paper trace—they move around in virtual space and enter into digital relationships that are specific to the computer. The rest of this book is devoted to showing precisely how data are constrained by the physical and virtual structures that create and store them.

3 Organizing Space

What do the spaces in which bioinformatic knowledge is produced look like? How are they arranged? How do people move around in them? What difference does this make to the knowledge that is produced? The dynamics of data exchange have driven a spatial reorganization of biological work. That is, data work demands that people and laboratories be arranged and organized in specific ways. The ways in which walls, hallways, offices, and benches are arranged, and the ways in which people move among them, are crucial in certifying and authorizing bioinformatic knowledge—the movement of data through space and among people renders them more or less valuable, more or less plausible. Because of this, spatial motion is also bound up with struggles between different kinds of work and the value of different kinds of knowledge in contemporary biology.

The volume and speed associated with high-throughput techniques have transformed biologists' knowledge-making practices. But computers have also necessitated reorganizations of workers and space that transform what counts as useful knowledge and work for biology. In particular, laboratory work itself is managed as data: samples, movements, and people are recorded in databases. These new practices have led to the quantification and control of space and work. Bioinformatics entails not

just speeding up, but rather making speed and efficiency virtues of biological work.

Biological data are valuable—they can be shared, traded, exchanged,
and even bought and sold under appropriate circumstances.[1] They are
a form of capital for performing biological work. Before the Human
Genome Project began, Walter Gilbert quipped that genome sequencing
would not be science, but production. But biology has become production. A large proportion of contemporary biology (particularly genomics) turns on the production of a product—namely, data. The machines
of this production are computers. The data orientation of contemporary biology should also be understood as an orientation toward productivity. Data have generated new modes of value in biological work.
Gilbert no doubt meant to denigrate work on the human genome as
mere production—mindless, uncreative activity that stood in contrast
to the work of proper science. But the pejorative connotation now rings
hollow—successful production requires high technology, intricate management, difficult problem solving. It is the products of these processes
that will comprise the raw material in the manufacturing of new drugs
and new medical treatments. Doing "good" work in bioinformatics is
now doing productive work—that is, work that contributes to the rapid
and effective creation and organization of data.

The link between space and the value of different kinds of biological work was already visible in the HGP. A significant part of the debate about the project centered on whether to distribute resources for
sequencing widely or to focus on a few large, centralized genome sequencing centers. Many feared that the HGP would compete with, and
ultimately reduce resources for, the type of small-scale biology that had
given rise to the most important discoveries.[2] At the heart of concerns
over the centralized approach was the suspicion that the HGP did not
really amount to science. It was feared that a "technological" focus
would have a pronounced influence on the future direction of biological
work, teaching young biologists techniques, but not creativity.[3] These
debates were simultaneously a contest over the organization of biology
in space and the *value* of certain kinds of biological work. In particular,
what was in question was the value of the "technological" work of using, producing, storing, managing, and sharing data.

In the first part of this chapter, I describe a division in bioinformatics between "producers" and "consumers" of data. Production and
consumption form a cycle, both parts of which are crucial for the generation of biological knowledge, but data producers are relegated to a
lower status. I show how these divisions of labor are inscribed in the

spaces of biological work and how biological knowledge is authorized through its movement between spaces. The last part of the chapter examines the production of data in more detail. It shows that production sequencing requires the kinds of efficient and accountable work that computers enable.

In describing Robert Boyle's "House of Experiment," Steven Shapin has shown how the physical spaces (and the flow of people, materials, and data within them) crucially structured the knowledge that was produced: "The siting of knowledge-making practices contributed toward a practical solution of epistemological problems."[4] The organization and policing of space, particularly the division into private and public, was central to the validation and authorization of knowledge claims. Despite the manifest differences between seventeenth-century houses and modern laboratories, something similar can be said about bioinformatics: the ways in which labs are organized and the ways in which data flow within them contribute to the solution of the practical epistemological problems posed by computer scientists, biologists, and database managers having to work together. The biological knowledge, the forms of labor, and the spaces of bioinformatics reproduce one another. The kinds of epistemic changes described in chapter 2 require the simultaneous transformations in work and space that are detailed here.

Divisions of Labor: Producers and Consumers

Attention to the dynamics of contemporary biology shows that there are two distinct kinds of work being carried on. First, there are individuals working to transform samples into data; second, other individuals analyze these data and make them into biological knowledge. These linked processes can be described as the production and consumption of data, and it is through a cycle of production and consumption that bioinformatic knowledge is generated. Both production and consumption require both biological and computational expertise. By marking consumption of data as high-value work, while considering data production to have lower value, biologists are able to police the kinds of knowledge that are considered valuable and acceptable in bioinformatics. This policing can be understood as a form of continued resistance to (and discomfort with) computational and data-driven approaches; data-driven biology has required biologists to defend their skills and expertise within a landscape of new problem-solving approaches. By labeling production as less "scientific" or "biological," biologists are able to maintain control of biological work.

In my interview with one biologist who had spent significant amounts of time in industry, he spoke at length about how he understood the difference between different sorts of workers:

> IT people are really like a dime a dozen . . . they are interchangeable parts for most large corporations; but a stats guy you just can't do that with . . . statistics is hard, computers are not . . . I experienced this firsthand at the Whitehead. . . . There would be these guys that would get hired in and they would get fired within a month because they were just programmers. And they would call themselves bioinformaticians. So we started to make a distinction at the Whitehead between bioinformaticians and computational biologists. . . . If you went to the Whitehead or the Broad today and said: "I'm a computational biologist," you'd have to prove yourself. And to prove yourself you'd be writing programs to do linear regressions models or peak finding for something, and if you're a bioinformatician you'd be running the Oracle database. It's like saying "I'm a computer scientist" versus a "I'm a computer engineer." [At the Broad] we were having a problem where we would hire people who said they were bioinformaticians and we would ask them to do something statistical or mathematical and they would just fall on their faces. . . . We would ask them to tell us what a Z-score means and they would be like: "uh, duh, I don't know." How do you call yourself a bioinformatician then? "Well I know how to run a database and query GenBank and get some sequences out." Yeah, everybody on the planet can do that! We would be like, "What do you mean, like type the sequence into the command line and get the file back and parse it with a parser you wrote ten years ago?" Yeah. Oh, that's a thing people do? . . . Millennium guys would be like "I'm a bioinformatician." Okay, Millennium Guy, what did you do? "I ran all the sequencing pipelines to look at transcripts." "Okay, cool, so did you use hidden Markov models or Bayesian analysis to do your clustering?" And they'd be like "Huh?" So we called bioinformaticians something different than industry was. And then the buckets became very clear: people who didn't have math or statistics backgrounds, those were the bioinformaticians; people who had math and statistics backgrounds, who were so important to the organization, those were your computational biologists. What

happened . . . was that you started to lower the salary that you would pay to the bioinformaticians, because those people are commodities; any failed startup company, you could go and find anybody who could do this stuff. Anybody with a statistics background, those people you can't find anywhere on a corner . . . so those people you would put a premium on. . . . [Bioinformaticians] are just a tool . . . you're a tool, you're a shovel . . . go put together a database of this many sequences and make it run on the BLAST-farm while I go and do the tweaking of BLAST.

Although not all biologists saw the divisions quite this starkly, an observable difference certainly existed between the so-called bioinformaticians and the so-called computational biologists. Although they both spent the vast majority of their time in front of computers, the two groups usually occupied different workspaces within the institutes and laboratories, ate lunch separately, and went to different parties.[5]

As is suggested in the quotation above, however, the two groups are also performing different sorts of tasks. For the one group, those tasks were often described as "informatic": building and maintaining databases, running BLAST queries on sequences, running data through a pipeline of ready-made software tools, or perhaps building and maintaining a website of biological data. "Computational" tasks, on the other hand, included data analysis with sophisticated statistics, designing and improving algorithms for sequence analysis, and implementing sophisticated mathematical models for reducing biological data.[6] It will be useful to provide one detailed example of each from my fieldwork:

: : :

Informatics: Helen is working at her computer. A biologist from the wet lab enters her office and explains an experiment he is doing on epitope tags[7] in a particular species of bacteria. The biologist needs Helen to find out what epitopes have been characterized for this organism and provide him with a list. This is a difficult task because it involves pulling information from published papers and potentially from databases. After several days of research scouring journals and the web, Helen puts together a list of online databases containing epitope tags. Being a skilled programmer, rather than retrieving the information by hand, she writes a program to pull the information off the web from a variety of databases, put it into a standard format, and generate her own database with all the information the wet-lab biologist needs.

Computational: As part of the work for a paper he is writing, Sam, a graduate student, needs to know the relative levels at which a particular gene is expressed in different tissues of the body. Sam organizes a collaboration with a nearby lab, which, after some months, provides him with the data he needs, drawn from experiments it has been doing on different cell lines. In the meantime, Sam develops a theoretical model of gene expression that he codes into an algorithm; it takes several days to run on the computer cluster to which he has access through his lab. When Sam receives the experimental data, he writes a different program, first to parse the data into a form he can work with, and then to perform a series of statistical checks: he knows he must be careful to control for differences between the experiments and differences in overall expression levels. Sam now has some results that he can compare with his own theoretical model of gene expression. This comparison will form the basis of a paper published in a well-respected biology journal.

The attitude toward "bioinformatics" expressed in the quotation above suggests that there is a strong hierarchy implied in the distinction between "bioinformatics" and "computational biology." This attitude— that data production is an unimportant sideshow to the real activity of biology—was held by many (although not all) of my informants. Moreover, this attitude has been internalized by many practicing bioinformaticians, who see their roles as a "service" for biologists. During my fieldwork at the Broad Sequencing Center, the bioinformatics team understood their role as providing "informatic support" for the core work of the biologists in their labs. Although some resented the fact that biologists would often demand that they dig up data or reorganize a database at a moment's notice, they saw it as an inevitable part of their day-to-day work in the laboratory. Like the stereotypical IT support personnel in a large corporation, bioinformaticians saw themselves as underappreciated and as playing an unacknowledged role in the success of the work being done. Some others became disillusioned with their "second-class" status: one of my interviewees had just quit his job at the Broad because he felt that his skills were undervalued. He had found a position at Google, where he felt that his expertise in informatics and managing large data sets would be rewarded both financially and in terms of prestige.

This hierarchy is built into the structure of the Broad itself, which is divided into "programs" (such as the Cancer Program, the Psychiatric Disease Program, and the Program in Medical and Population Genetics) and "platforms" (such as the Genome Sequencing Platform, the Imaging Platform, and the Biological Samples Platform). Broad personnel generally understand the programs to be performing the real research work of the institute, while the platforms provide support in the form of imaging, cheminformatics, and sequencing. The Computational Biology and Bioinformatics program recruits mostly individuals with strong mathematics and statistics backgrounds, while the lower-status "bioinformatic" work is distributed among the supporting platforms.[8]

We should not be too hasty in following practitioners in dividing bioinformatics into "higher" and "lower" forms—both contribute to knowledge production in crucial ways. The attitude that informatics is less valuable to an organization than computational biology is contingent upon a particular notion of "value" for biology. The private sector (represented by "Millennium Guy") holds a different view of what count as valuable skills for biological work. For pharmaceutical companies like Millennium, value is generated by "high throughput": finding potential drug targets by testing thousands of proteins; finding a chemical with drug potential by screening hundreds of chemicals; or predicting potential toxic effects by checking thousands of possible "off-target effects." Such tasks do not necessarily require a detailed understanding of how a particular biological pathway or system works; since the space of potential compounds and potential targets is immense, speed and volume are crucial.

But this kind of bioinformatics is not only valuable within the private sector: these kinds of practices are also highly regarded *within* specific types of biological work. For instance, in the Broad Sequencing Center, the aim is to produce as much sequence as quickly and cheaply as possible with the smallest number of errors. Workers who can design better databases, streamline sequencing pipelines, and find ways to process sequence data more quickly are highly sought after. Brendan, a programmer, told me how he designed a Laboratory Information Management (LIM) system—an internal lab database—that greatly improved the sequencing process. The SQUID-BetaLIMS system, which was still used by the Broad during my fieldwork, allowed the lab to track samples, diagnose errors, and track workflow in ways that were crucial to its performance as a state-of-the-art sequencing laboratory.

Although Brendan had an undergraduate degree in biochemistry, he had worked as a computer programmer for twenty-five years—he assumed that his "domain knowledge" of biology had very little to do with his being hired. Rather, it was his software and database expertise that was exactly what was required for this kind of biological work.

The pronouncement that "bioinformatics" occupies an under-laborer status within biology is designed to suppress the importance of certain new kinds of biological practices. Indeed, these practices are portrayed as "unbiological"—they are, some biologists insist, just informatics or just engineering or just management, and therefore distinct from the real business of doing biology. Such workers, and the spaces in which they work, should be removed, as far as is possible, from view.[9] What is at stake is perceptions about what biological practice *should look like*. When Gilbert made his claim that sequencing was not science, but production, he was right. But that fact has made many biologists profoundly uncomfortable; biology, according to many, means making hypotheses, doing experiments, and drawing conclusions about living systems (whether this means using computers or not). It does not mean building databases, managing information, and reorganizing laboratory workflow. Yet these latter practices are now as much a part of biological practice as the former. What we see in bioinformatics is biology trying to come to terms with these different forms of work and knowledge making. As attributions of value, the terms "bioinformatics" and "computational biology" are weapons in a battle over what kinds of practices will count as legitimate biology in the years to come.

Spaces of Work

In the rest of this chapter, I will show how these divisions of labor are generated and maintained by the physical spaces in which these new forms of biology take place. The organization of space manages the tension between different forms of knowledge work. Data-driven biology requires not only specific divisions of labor, but also specific kinds of spaces and movements within them. The "higher" (consumption) and "lower" (production) forms of practice are accorded different sorts of spaces and involve different sorts of physical movements. The architecture and space within which the knowledge is produced contributes to its epistemic value. Specific movements and configurations of people, objects, and data act to verify and certify biological knowledge in various ways. This section will draw primarily on fieldwork that took place during 2008 at the Broad Institute in Cambridge, Massachusetts.

FIGURE 3.1 Front exterior view of the Broad Institute at Seven Cambridge Center, Kendall Square, Cambridge, Massachusetts. (Photograph by author, November 2010.)

The Broad Institute, endowed by $100 million gifts from Edythe and Eli Broad, Harvard, and MIT, lies in the heart of Kendall Square. A block from Technology Square and standing directly across Main Street from the MIT biology building and the lopsided ridges of Frank Gehry's Stata Center, the building at Seven Cambridge Center is the epitome of twenty-first-century laboratory chic. "7CC," as those who work there know it, is eight stories of shimmering glass and metal straddling almost the entire block (figure 3.1). The vast double-story lobby houses a custom-designed information center assembled from hundreds of flat-screen televisions and clearly visible from the other side of the street. A small "museum" of sequencing instruments, plush leather couches, and a serpentine glass staircase, more appropriate to a Californian mansion than a laboratory, fill the remainder of the space. The upper floors of the lab are accessible via a bank of elevators, which can be activated only by an RFID card. Although some of floors have been outfitted for specialized purposes (for instance, to house the large array of servers or the chemical screening robots), the basic pattern of many of the floors is identical. On the east side, and occupying about two-thirds of the floor space, are offices and conference rooms. Almost all of these have glass walls, allowing light to penetrate from the exterior windows into the central area of the building. On the west side of the building, separated from the offices only by glass doors and walls, are the laboratory spaces proper.

On the one side, people sit at their computer terminals; on the other, they stand at bench tops, pipetting or carefully adjusting various medium-sized instruments. From almost anywhere on either side, it is possible to see directly into the other, even though, for reasons of biological safety, doors must remain closed and air pressures must be maintained at different levels. In the open spaces around and between the offices, the overall impression is one of openness—ample light, neutral tones, glass, and the white walls and benches of the laboratory create a space that feels both scientific and businesslike, both a lab and a management consultant's office. For most visitors, scientific and otherwise, this is the Broad Institute. However, no DNA sequencing—the activity on which almost all other activities at the Broad depend—takes places at 7CC. Less than a ten-minute walk away lies the building known as "320 Charles," or the Broad Sequencing Center. Although only a few blocks distant from 7CC, 320 Charles presents a very different scene. On the other side of Binney Street, behind AstraZeneca, BioGen Idec, and Helicos Biosciences, this part of East Cambridge reminds the observer of the abandoned and dilapidated mill towns of western Massachusetts. The neighbors of 320 Charles include an outstation for an electric company and a pipe manufacturer. The building itself shows almost no external signs of the high-tech science within—only a small plaque near the entrance emblazoned with the Broad insignia would provide a clue to a passerby. Also a block long, 320 Charles is low and squat, barely two stories high. Its only external features are large ventilation shafts on the roof and warehouse-style truck-loading docks at the rear (figure 3.2). Indeed, the building used to be used by Fenway Park as a warehouse for Red Sox paraphernalia, and for a time, the sequencers and the famous red and white jerseys must have shared the space before the Broad expanded and took it over in its entirety.

Inside, the warehouse feeling is still evident. The interior, presumably fairly rapidly adapted to its new purpose, still seems somewhat uncomfortable in its role. In contrast to the open, easily navigated spaces of 7CC, 320 Charles presents itself to the newcomer as an almost impenetrable rabbit warren. Windowless rooms, long corridors that seem to lead nowhere, unevenly partitioned spaces, unnecessary doorways, and flights of stairs leading to odd mezzanine levels are a result of the several rearrangements that the lab has undergone as it has had to be adapted to suit various needs. When I first took up my place in the lab, it was

FIGURE 3.2 Exterior view from street of Broad Sequencing Center, 320 Charles Street, Cambridge, Massachusetts. (Photograph by author, April 2008.)

several weeks before I summoned the courage to enter by a door other than the one immediately adjacent to my office for fear of becoming embarrassingly lost in the interior. Rubber flooring, uniformly white walls, and metal staircases add to the provisional feel of the place. If 7CC is a consulting office, 320 Charles reminds one of nothing more than a manufacturing plant. Although there is office space, it is pushed toward the edges of the building. The largest, and most central, room in the building is the cavernous space housing the sequencing machines. Large ventilation pipes fitted across the length of the room maintain air quality and temperature at precisely monitored levels. The only noticeable motion comes from an electronic ticker-tape display on the back wall that scrolls the current average length of the sequencing runs in red and green letters.

The two main buildings of the Broad Institute represent different forms of biological practice and serve different scientific ends. It is useful here to draw on the sociology of Erving Goffman. Goffman's analysis of "region behavior" draws a distinction between the "front region," in which a social performance is given, and the "back region," where the performer can "drop his front."[10] It is in the back region that a good deal of hidden work is performed—work that is necessary for maintaining the performance taking place in the front region.[11] As Goffman notes,

> it is apparent that the backstage character of certain places is built into them in a material way, and that relative to adjacent areas these places are inescapably back regions. . . . Employers complete the harmony for us by hiring persons with undesirable visual attributes for back region work, placing persons who "make a good impression" in the front regions.[12]

While I by no means wish to suggest that workers at 320 Charles have "undesirable visual attributes," there are many ways in which the relationship between the two Broad buildings maps onto the division between Goffman's front and back regions. First, the locations of the two buildings make their relative status clear. 7CC is a front region, designed for showing off biological work to scientific and nonscientific visitors. As a showpiece of Kendall Square, its grand, glassy lobby is designed to create and sell an image of biological work. 320 Charles, on the other hand, is sequestered in an industrial zone. The technical work of this back region sustains the performance of 7CC.

Second, 7CC and 320 Charles differ in terms of access. The former appears open and welcoming to the visitor—anyone can wander into the lobby to peruse the sequencing "museum" or the televisual display. Although access to the upper floors is controlled, scientific visitors are afforded office space and free movement about the informal meeting spaces, kitchens, balcony garden, and library. 320 Charles, on the other hand, appears almost totally closed off: there are few exterior windows, access to any part of the building is by ID card only, doors are monitored by security cameras, and the interior space is closed off and hard to navigate. Biological laboratory safety designations (BL1, BL2, BL2+) further designate spaces that only individuals with certain levels of training are permitted to enter. Third, human traffic between the two laboratories is controlled—few individuals find the need to regularly make the ten-minute walk from one building to the other. In this way, 320 Charles remains remarkably hidden and cut off from its counterpart.[13]

Robert Tullis, one of the principal architects for the 7CC building, told me that Eric Lander wanted 7CC to be a space in which many different types of people could be comfortable: undergraduate and graduate students, senior faculty scientists, software engineers, consultants, and visitors from the NIH and NSF. Lander wanted a "sock track" around each floor of the building—a "racetrack corridor" that one

could traverse with one's shoes off and still be able to see every aspect of
the laboratory at work (of course, no one would be allowed in the wet
lab spaces proper without shoes). This pathway was designed so that
Lander and other lab leaders would be able to show visitors around the
lab—to show off the biology that was being performed there to other
scientists and to potential donors and funders. Originally the building
at 7CC had been conceived as space for retail. However, Lander saw the
open spaces and large amounts of glass as an opportunity to create a
highly transparent laboratory space. "Eric Lander's instructions to us,"
Tullis recalled, "were that he wanted [the lab] to be about transparency:
clean, bright, and open. It was a philosophical desire. Since the purpose
was to do genomic studies, and make results available, the transparency
of discovery should make its way into the architecture."[14] As a place of
both science and business, an amalgam between office and laboratory,
7CC—the cleanliness and order, the light and neutral tones, murmured
conversations in the hallways—reminds the observer of a physician's
office.

The division between the front region of 7CC and the back region of
320 Charles inscribes a division between two distinct kinds of practice
and two distinct products. The latter building employs many individuals
performing repetitive lab bench work or managing databases. The output is large amounts of biological data. 7CC, on the other hand, houses
mostly PhD scientists performing statistical and mathematical analyses,
writing software, and performing experiments. Their work is mostly
to produce scientific papers based on the analysis of large amounts of
data. Each of these two distinct regimes of biological knowledge is represented by the physical spaces in which it is performed.

During the design of 7CC, it was realized that the Broad program
had expanded such that the number of sequencing machines required
would exceed the number that it would be possible to house on site.
The solution was 320 Charles: the conversion of the warehouse would
provide adequate square footage for the large number of sequencers
required. DNA sequencing is a highly repetitive activity, and the facto-
rylike layout of 320 Charles is appropriate to the performance of such
labor. Indeed, the appearance and design of the building assure the visitor that it is the sort of place suited to outputting large volumes of iden-
tically replicated products (like the Red Sox jerseys that were originally
stored there). Access is limited to those few who are needed to run and
manage the day-and-night output of data. The central place given to the
sequencing machines symbolizes the notion that 320 Charles would ideally run as an automaton. Careful attention is paid to the specific layout

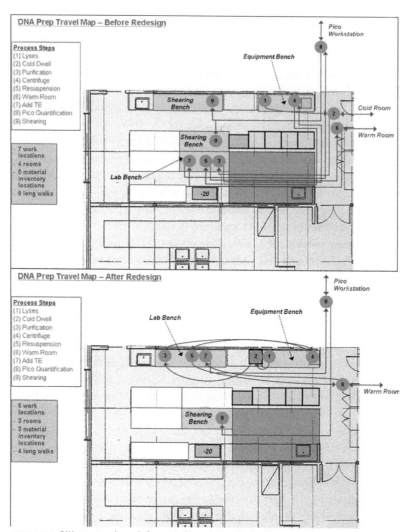

FIGURE 3.3 DNA prep travel map before and after redesign. The locations at which each step is performed are indicated by the circled numbers. Movement between locations is shown by lines and arrows. The "after" map shows a reorganization of the benches and work locations and a reduction of movement between locations. (Vokoun, "Operations capability improvement." Copyright Massachusetts Institute of Technology. Reproduced with permission.)

of lab benches and essential equipment so that the lab can run, if not as a machine, then as a Taylorist production line in which every worker executes precise and repetitive motions. Figure 3.3 shows a workflow diagram produced by a student at MIT's Sloan School of Management in order to streamline productivity. The ultimate measure of productivity is the error rate, which is carefully monitored by a team of quality control engineers. The appearance and the configuration of the space, and the movement within it, certifies the knowledge that is produced.

The division between the front region of 7CC and the back region of 320 Charles largely matches the division between "higher" and "lower" bioinformatics described earlier in this chapter. The difference in status between biological practices based on mathematics, statistics, and bench work and biological practices based on informatics and databases is represented in the physical spaces of the laboratories in which they are performed. The notion that the former practices constitute the "real" work of biology and that production sequencing is mere technical support is reinforced by their front region/back region configuration. Selling biology to its funders, as Lander is well aware, means projecting an image (putting on a performance, if we are to follow Goffman's terminology) of a biology that not only is open and transparent, but also offers immediate and tangible benefits for human health. In September 2008, the Broad Institute received an additional (and record-breaking) $400 million gift from Eli and Edythe Broad: "Of all our philanthropy," Eli Broad announced, "the Broad Institute has been the investment that has yielded the greatest returns. . . . We are convinced that the genomics and biomedical work being conducted here . . . will ultimately lead to the cure and even the prevention of diseases."[15] 7CC is highly conformable to the image of a place in which such medical advancements will be made; its configuration not as a traditional laboratory but as a mixed lab-office space associates it with medical practice.

At the Broad, this strong emphasis on medical applications requires a fluid and immediate interaction between biologists and computer scientists. A typical problem might be to determine all the locations in the human genome that are strongly linked to the development of Crohn's disease.[16] Such work involves gathering samples from a large number of patients (usually at an affiliated hospital), processing the samples, performing the genotyping for each sample, developing computational and statistical methods to analyze the entire set of genotypes, and finally examining the results of this analysis to see whether they provide any clues about the biological pathways through which the disease acts. Close and rapid communication between those collecting the samples (medical doctors), those preparing them (wet-lab biologists), and those performing the analysis (computer scientists) speeds the process. 7CC presents a vision of integrated, medically oriented, scientifically rigorous, and productive biology. The work done here is perceived as valuable by those both inside and outside the Broad because of its applicability to medicine. This applicability is achieved at least in part by creating a space that is designed and operated to produce this value.

Thomas Gieryn has commented that "built places materialize identities for the people, organizations, and practices they house. Through their very existence, outward appearances, and internal arrangements of space, research buildings give meanings to science, scientists, disciplines, and universities."[17] This is an apt description of the two buildings comprising the Broad Institute. The segregation and spatial organization of different kinds of practices promotes a particular vision of biological work. First, it renders some practices largely invisible, deeming them beyond the boundaries of "real" biological practice. Second, the architecture of 7CC denotes a vision of collaboration between biologists and computer scientists: through the mixing of their spaces into one another, the observer is given the impression that the two different sorts of practices are collaborating harmoniously. Gieryn argues that people are designed along with the walls and windows—that spaces give people identities as particular kinds of workers and particular kinds of knowledge producers. The spaces of 320 Charles identify those that work there as technicians, data producers, and shop-floor managers; the design of 7CC suggests academic science, knowledge work, and even medicine.

Lander's notion of the "transparency of discovery" means a constant sharing and movement between wet and dry spaces at 7CC. "Our research only works in the spirit of collaboration," Alan Fein (a deputy director of the Broad) comments. "Everybody needs to be accessible, transparent, visible."[18] If everyone can see what everyone else is doing, they will be able to understand one another's work and collaborate most effectively, the designers hope. This is a vision of a new, integrated, hyper-productive biology. It is also a highly sellable biology.

The Lab That Changed the World

But the division into higher and lower, front and back, obscures the full range of identities and practices that this biology relies on. Bioinformatics depends just as much on the productions of 320 Charles as it does on the work of 7CC. Let's now examine the back region of the Broad in order to describe what goes on there in more detail. What we will discover is that new modes of valuable work are emerging in this sort of activity. Here, principles of management and efficient manufacturing have been rigorously applied to the making of biological data. This is certainly production, but it is not mere production: it requires careful attention to the organization of space, people, and technology. Ultimately, these spaces are structured and monitored by computers; samples and people are represented there as data too.

When I visited the Broad Sequencing Center at 320 Charles, one of the first people I spoke to was Meredith, the manager of the Molecular Biology Production Group. Her office sat on a floor above and overlooking the sequencing lab, and as I traveled along the hallway I peered down on the workers busy at their lab benches. The first thing I noticed in Meredith's office was that the bookshelves were almost empty except for about fifteen copies of a single book: *The Machine That Changed the World: The Story of Lean Production* (1991). I asked the obvious question: Why all the books? "It's required reading for my employees," she told me, "every new person on my team gets a copy." Perhaps surprisingly, this isn't a book about molecular biology, or about any natural science, but about assembly lines.

The tagline of *The Machine That Changed the World* is "How Japan's secret weapon in the global auto wars will revolutionize Western industry." The book is based on a detailed study of the Japanese automobile industry by three of the directors of the International Motor Vehicle Program at MIT, James Womack, Daniel T. Jones, and Daniel Roos. "Lean production" (in contrast to Henry Ford's "mass production") is the name they give to the techniques deployed in the Japanese car industry (developed largely by Eiji Toyoda and Taiichi Ohno) in order to manufacture high-quality products at a low cost.

> The craft producer uses highly skilled workers and simple but flexible tools to make exactly what the consumer asks for—one item at a time. . . . The mass producer uses narrowly skilled professionals to design products made by unskilled or semiskilled workers tending expensive, single-purpose machines. These churn out standardized products in very high volume. . . . The lean producer, by contrast, combines the advantages of craft and mass production, while avoiding the high cost of the former and the rigidity of the latter. Toward this end, lean producers employ teams of multiskilled workers at all levels of the organization and use highly flexible, increasingly automated machines to produce volumes of products in enormous variety.[19]

Because of the expense of equipment and its intolerance of disruption, mass production tends toward oversupply of workers, space, and raw materials; workers are bored, and there is little variety in products. Lean production, on the other hand, has the potential to reduce human ef-

fort, reduce space, reduce inventory, reduce engineering time, and produce greater variety. One of its goals is also to "push responsibility far
down the organizational ladder," making workers able to control their
own work. Indeed, lean production relies on "an extremely skilled and
a highly motivated workforce," in which employees become members
of a "community" that must make continuous use of its knowledge and
experience. In the Toyota plants, for instance, Ohno placed a cord above
every workstation, which workers could use to stop the entire assembly
line if they encountered a problem they could not fix; this was in stark
contrast to a mass production line, which could be stopped only by
senior line managers in special circumstances.[20] Workers were encouraged to identify and rectify the cause of the problem. Lean production
depends not only on teamwork, but also on proactive problem solving
by every worker.

In keeping with the spirit of valuing workers and their work, lean
production depends on close coordination and cooperation between design engineering and production. At Honda, university-trained mechanical, electrical, and materials engineers spend their first three months
of work assembling cars on the production line.[21] This not only makes
the engineers aware of how different aspects of the business work, but
also fosters teamwork and communication. It makes engineers acutely
aware of the problems that their designs are likely to encounter on the
assembly lines, so that they can anticipate or avoid those problems in
their design work.

Lean production also depends on having small inventories of supplies (sometimes called the just-in-time system), which saves on storage
space, and close relationships with suppliers, which allows sharing of
information about products. Having a small inventory means "working
without a safety net"—if there is any defect in a component from a supplier, production will be disrupted. This difficultly is mitigated by what
Toyota calls the "five whys" (presumably: "Why? Why? Why? Why?
Why?): "Both the supplier and the assembler are determined to trace
every defective part to its ultimate cause and to ensure that a solution is
devised that prevents this from ever happening again."[22] Problems with
components are solved rather than absorbed into the total cost or time
of production.

The Broad Institute, and particularly its sequencing operations, had
a commercial tenor from its beginning. Its founder and director, Eric
Lander, had taught managerial economics at Harvard Business School
before founding the Whitehead Institute/MIT Center for Genome Research in 1990. Robert Nicol, the director of the high-throughput Ge-

nome Sequencing Platform, a chemical engineer by training, had worked previously as a project manager for the Fluor Corporation, "the largest US-based, publicly traded engineering and construction firm." Rob came to MIT in 1999 as a fellow in the Leaders for Manufacturing Program to conduct research on manufacturing systems and processes, joining the Whitehead in 2001 in order to implement industrial process design, control, and improvement techniques.[23]

During the time I spent at the Broad Sequencing Center, the influence of Nicol's manufacturing and industrial process design ethos could be seen everywhere. Not only lean production, but Six Sigma (6σ) and a range of other manufacturing techniques had been put into practice in biology.[24] First, a large amount of time and effort had been invested in planning and streamlining processes and workflow for sequencing. Space and materials were carefully organized to economize human and sample movement around the labs. Beginning in 2003, the Broad recruited a series of MBA students from the Sloan School of Management (MIT's business school) to investigate the potential for improving the manufacturing capabilities of the lab. As Matthew Vokoun reported, "During the completion of the HGP in the 1990s and early 2000s, the purpose of the Broad Institute's sequencing operation was to rapidly scale-up or 'industrialize' the genome sequencing process. This industrialization of sequencing refers to its transition from being a highly skilled craft performed by a few very well-educated biologists to a large-scale, coordinated production process involving over one hundred technicians, engineers, managers, and scientists."[25] This was achieved by breaking the sequencing process down into small, repetitive steps that could be performed quickly and accurately.

In 2003, Julia Chang was given the task of analyzing and improving the process of "picking" E. coli colonies for sequencing.[26] Picking is an automated process through which E. coli colonies growing on agar plates—each containing a distinct fragment of DNA—are transferred to 384-well plates for sequencing. Colonies used to be picked by hand using a toothpick. At the Broad, a digital camera images the agar, and a specialized software program analyzes the image to determine the positions of colonies with desirable characteristics (size, roundness, distance from other colonies). A computer arm fitted with specialized tips then transfers the suitable colonies to the wells. Chang's mandate was to identify sources of variation in this process and suggest steps to eliminate them. Working with the team responsible for picking, Chang devised a series of experiments to determine what variables most influenced the yield of colonies successfully transferred to wells. Chang used her experience with operations management and process control theory to analyze the results.

One of Chang's key findings was that significant variability in yield was caused by variation in the density of the colonies grown on the agar plate. This observation led to the development of a new process for plating the *E. coli* on the agar using microfluidic techniques, which eliminated the inherent variation in the number of cells transferred to the plates with each dispensed volume. As Chang noted in the conclusion to her thesis, "While not widely available or referenced by those in the organization, sufficient paper records contained the data required to build control charts of the picking process. The documented variability seemed typical of traditional industrial operations and suggested that operational methodologies would have some traction."[27] In other words, Chang collected data that had not been considered relevant or interesting to the Broad's technicians and mobilized them to formulate new and more productive sequencing practices.

The following year, Vokoun, who had worked previously as a process development engineer in the Optical Systems Division at 3M, attempted to apply operations management techniques to the Molecular Biology Production Group (MBPG).[28] As the "most upstream" part of the sequencing process, the MBPG was the least automated and most "craft"-dependent part of the lab. The aim of Vokoun's work was to transform the MBPG's "highly variable output" by implementing lean production, production forecasting, Six Sigma, and RFID (radio-frequency identification). Beginning in July 2004, Vokoun managed a five-month lean production implementation project in MBPG with five goals: (1) eliminating all chances of mixing up DNA samples; (2) creating personal workstations with full sets of equipment and materials; (3) minimizing travel for samples and workers; (4) improving and standardizing materials flow; and (5) cleaning up and organizing the MBPG area, recovering unused space.[29]

These changes were based on several principles of lean production, including 5S, pull production, and *kanban*. 5S, from the Japanese words *seiri, seiton, seiso, seiketsu*, and *shitsuke* (translated as "sort," "straighten," "shine," "standardize," and "sustain"), is a method for organizing workplaces and keeping them clean. Pull production also refers to a method of organizing workstations by simplifying material flow through the workspace. Workstations constitute a "sophisticated socio-technical system" in which there is "minimal wasted motion, which refers to any unnecessary time and effort required to assemble a product. Excessive twists or turns, uncomfortable reaches or pickups, and unnecessary walking are all components of wasted motion."[30] *Kanban* is translated from Japanese as "visible record"—it embodies the principle that the flow of materials must be carefully managed in order to limit inventory in the pipeline.

Vokoun used 5S, pull production, and *kanban* to recreate the modes of technical production within the MBPG. Working closely with the technicians, Vokoun gained hands-on experience with a particular step of the sequencing process known as "ligation." First, he sought to identify problems: process travel maps were drawn, cycle times were measured, and equipment lists were made. Figure 3.4 shows hand-drawn maps of the movement of workers around the lab during the ligation step. Vokoun's redesigned workflow reduced the manual time involved in the process from 9.3 hours to 6.1 hours.[31] Likewise, figure 3.5 shows photographs of the ligation workstations before and after their redesign according to the principles of 5S. The ligation team also created specialized kits containing all the reagents needed for the preparation of one DNA library, avoiding multiple trips to the storerooms or freezers. As can be seen from these figures, Vokoun's focus was on creating efficiencies by ordering space: moving equipment, economizing movement, creating visual cues, and making sure materials were where they could be best utilized. A similar redesign was undertaken for the transformation and DNA preparation steps, resulting in an overall redesign of the MBPG lab space. Vokoun concluded that the problems he encountered "had nothing to do with the actual molecular biology processes performed but rather were managed into the process by the policies, workflow designs, and organizational design of the MBPG."[32] What made the MBPG—and by extension, the Broad as a whole—successful or unsuccessful was not the quality of the "biology," as conventionally understood, but attention to the details of the operation as a manufacturing and industrial process.[33]

The requirements of sequencing operations also demanded new ways of organizing people. Scaling up to production sequencing meant hiring people with experience in engineering and in managing large teams. Knowing something about biology was important, but it was even more important to know how to organize people and how to manage projects. On the floor of the sequencing lab, PhDs in biology are few and far between. Many of the workers are young, often coming straight from undergraduate degrees in biology; there are also a disproportionate number of nonwhites and immigrants.[34] The tasks to be performed are often repetitive but depend on a high degree of skill and precision (for example, pipetting an identical, precise volume of solution over and

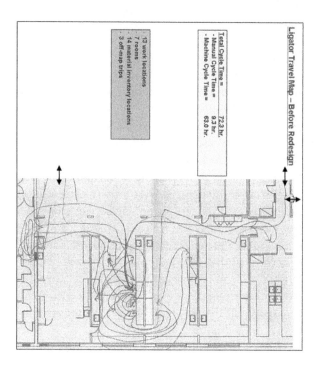

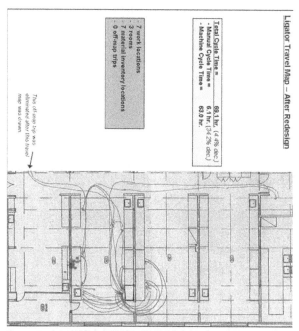

FIGURE 3.4 Ligator travel map before and after redesign. Hand-drawn arrows show workers' motion through the lab space. The "after" map shows a dramatic reduction in wasted motion across the lab corresponding to a reduced "cycle time." (Vokoun, "Operations capability improvement." Copyright Massachusetts Institute of Technology. Reproduced with permission.)

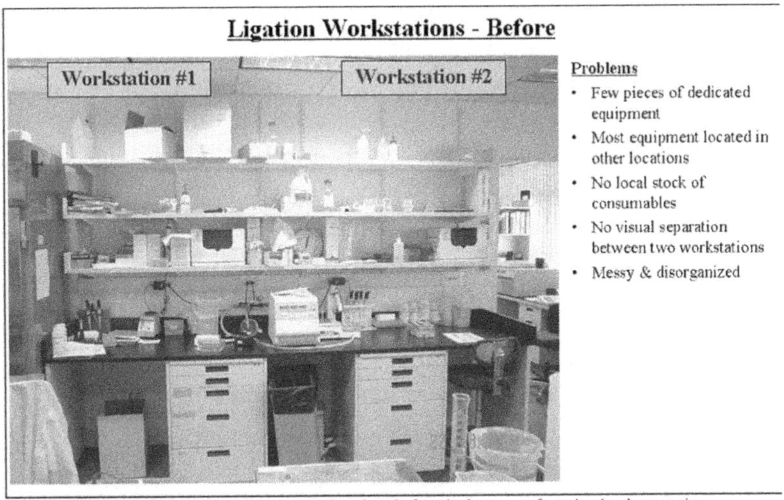

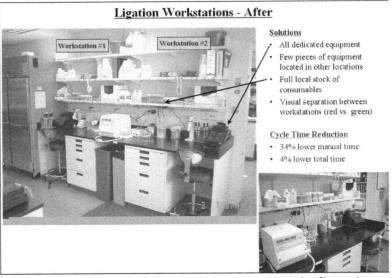

Ligation Workstations - Before

Workstation #1 **Workstation #2**

Problems
- Few pieces of dedicated equipment
- Most equipment located in other locations
- No local stock of consumables
- No visual separation between two workstations
- Messy & disorganized

Figure 29. Picture of two ligation workstations before the lean manufacturing implementation.

Ligation Workstations - After

Workstation #1 **Workstation #2**

Solutions
- All dedicated equipment
- Few pieces of equipment located in other locations
- Full local stock of consumables
- Visual separation between workstations (red vs. green)

Cycle Time Reduction
- 34% lower manual time
- 4% lower total time

FIGURE 3.5 Ligation workstations before and after redesign according to principles of lean production. The "after" image shows full stocking of necessary materials, organization of equipment, and visual separation between the two workstations. (Vokoun, "Operations capability improvement." Copyright Massachusetts Institute of Technology. Reproduced with permission.)

over). This circumstance—dependence on both the repetitiousness of mass production and the high skill of craft production—lends itself to the deployment of lean production.

Critically, workers are given a large amount of responsibility for organizing and improving their own work practices. For instance, every three months, workers in the MBPG are given a two-week "sabbatical" to reflect on their work and to come up with schemes for improving and streamlining the processes and workflows in which they are involved. Despite the repetitive nature of many tasks, managers realize that the success of projects ultimately depends on the skill and the commitment of individuals. One study of computer "finishers," for example, recognized the difference in interests between "workers" and "managers":

> The Center's senior management consisted primarily of academics and researchers, many of whom had pioneered modern gene sequencing. Typically PhDs, these managers held long-term career interests in the field of genomics. Though they ran a production facility, their ambitions also included publication, tenure, and senior roles in industry. This background contrasted sharply with that of the finishing personnel. Coordinators and finishers were typically young, in possession of a bachelor's degree, and at an early stage in their career. Some aspired to long-term careers in genomics or medicine. For others, finishing represented a temporary stopping point on the way to other careers.[35]

Making finishing more efficient meant rethinking incentives and reorganizing teams to bring management and worker goals into accord. The challenge was to maintain a sense of "pride and ownership" in the work while fostering cooperation and teamwork. Scott Rosenberg, an analyst from MIT's Leaders for Manufacturing Program, proposed new metrics for measuring finisher performance that would foster employee growth, encourage teamwork, and reward innovation as well as measuring individual performance.[36]

Moreover, Rosenberg proposed new ways to organize finishing teams in order to encourage collaboration and knowledge sharing. The difference between the original "skills-based" teaming, which assigned tasks to finishers on the basis of their experience, and the new "triage-based" teaming, which allowed junior finishers to try their hand at more difficult tasks, is illustrated in figure 3.6.[37] By allowing junior finishers

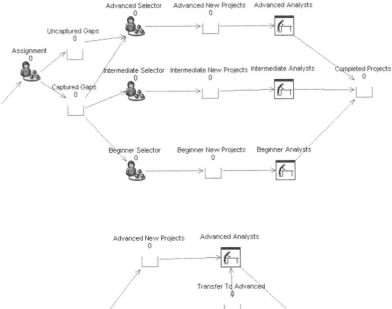

FIGURE 3.6 Skills-based and triage-based organization of finishing teams. In skills-based organization, (top) jobs are assigned according to skill level and are not passed between finishers at different levels. In triage-based (bottom) organization, jobs can move vertically between skill levels according to triage assessments. (Rosenberg, "Managing a data analysis production line." Copyright Massachusetts Institute of Technology. Reproduced with permission.)

to "hand off" their work to their more senior colleagues if they could not complete the task, triage promoted communication and knowledge sharing among all finishers. When finishing had been a small-group activity, Rosenberg realized, "its self-image tended to reflect the dedication and individuality of its members," who often worked nights and weekends to complete tasks. But such individual "heroics" were inappropriate and even counterproductive for a large production-line environment: "The organization was simply too large for its members to learn without the aid of better communication and collaboration."[38] Triage teaming provided a way to increase productivity while still recognizing and exploiting the special skills of individuals.

In order to encourage commitment to the organization, the Broad provides a career path in biology for individuals without PhDs. In particular, it fosters ways to move from the lab floor into supervisory, management, and planning positions. Several individuals that I interviewed had progressed to their present roles in this way, starting out mixing chemicals on the lab floor and eventually becoming responsible for large teams and the planning of workflows and sequencing processes. In 2008, Beth had worked at the Broad for seven years. After working as a health inspector in a local town, Beth had worked for the Massachusetts State Laboratory while earning a master's degree in biology from the Harvard Extension School. Her first job at the Broad had been "making reagents and solutions," and at first she had "no idea what DNA sequencing was." After several years, Beth worked her way up into the technology development team. By the time I spoke with her in early 2008, Beth had become a project manager in the quality assurance team, in charge of logistics, supply chain, and quality control for many of the materials coming into the lab. Likewise, Ben came to the Broad with a BA in biology, beginning his career mixing solutions in the materials lab. From there he moved to the "production floor" as part of the MBPG, and finally to the technology development group. Ben's role in that group was to scale up the processes for the new sequencing machines from the bench to the mass-production scale.

These examples demonstrate how the Broad operates reward systems outside the traditional academic channels of publication and tenure. Individuals who can work in teams, who exhibit aptitude for logical thinking and planning, who can design processes that bring efficiencies to the data production process, are promoted. Biological knowledge—especially of fundamental biological principles—is valuable, but it must be combined with an understanding of how a particular production process works and how it might be sped up by reorganizing materials or people. The success of the Broad Sequencing Center depends on a special kind of worker who is neither an automaton in the Fordist sense nor a lab bench scientist in the mode of a Pasteur or a Sanger. Instead, he or she (and both genders are well represented) is what might be called a "lean biologist," knowing only enough biology to perform the work efficiently. The lean biologist combines the individuality and creativity of the scientist with the work ethic and team orientation of the production line worker.

In addition to its careful organization of space, materials, and people, a final unique feature of the Broad was its orientation toward control. Keeping space, materials, and people in order means constant oversight. Meredith described an example of what she called the Broad's sophisticated sense of "operations." A few months before we spoke, certain sets of sequences had started to diminish in quality on the sequencing machines, producing shorter read lengths than average. At many labs such a problem would be (at worst) ignored or (at best) take months to resolve, leaving the sequencing machines running at suboptimal capacity. At the Broad, however, the monitoring was careful and sophisticated enough that the problem could be quickly traced to a particular batch of reagent from a particular outside supplier. The supplier was notified of the defects in the product and quickly supplied a new batch, and the problem was resolved within a few days.

Such a feat could be achieved through the tracking and monitoring of everything within the sequencing center. From the moment samples enter the lab (and often before), they are given a two-dimensional barcode that links the contents of the sample to its record in the laboratory database. As the sample moves through the lab, the barcode is scanned at every step: each machine in the lab is fitted with a barcode scanner so that the database can keep track of when each sample is run through that machine. Workers in the MBPG have scanners on their bench tops so that samples passing through their workspace are scanned in and out. Using the database, it would be possible to find the exact location of any given sample at any time; it would also be possible to find out which picking or sequencing machine it ran through, whose bench tops it passed over (and how long it spent there), and which batches of chemicals were used to treat it. All over the lab floor, large signs remind workers, "Never remove a barcode from anything!" The barcoding system is integral to the lab's ability to control its operations and monitor its workflow.

Indeed, the barcoding system is just the front end of a more thoroughgoing system of monitoring. That system goes by the names of quality control (QC) and quality assurance (QA). As a QA project manager, Beth was responsible for developing "Bills of Materials" (BoMs), detailed lists of quantities of materials used for each step of the sequencing process. By comparing the BoMs with sequence output, managers could call workers to account for the quantities of materials they were using. For instance, a Bill of Materials might allow 100 milliliters of ethanol and three pairs of rubber gloves per megabase sequenced; significant deviations from these quantities would quickly attract the attention of the quality control teams, who would investigate the discrepancies. Others in quality control designed tests to check the quality of both incoming reagents and outgoing products. Barcodes allowed a certain degree of oversight—one could compare, for instance, sequence read lengths from a reagent from supplier A with those from a similar reagent from supplier B, or the quality scores of data coming from samples prepared by worker A compared with those from worker B. But often this was not enough—in order to improve processes, "development" subteams in each sequencing team designed specific tests to measure the effects of using more or less reagent, or a cheaper alternative, or a faster process. They might ask, for instance: "Could we be using less TAQ polymerase and getting the same quality output?" These processes allowed the Broad to track workers on the lab floor, counting the number of pipette tips they discarded or the amount of a reagent they used in order to perform a particular sequencing step. If particular workers were found to use, for instance, fewer pipette tips for the same quality of product, a whole team could adopt their techniques. Little by little, the cost of the whole operation could be whittled down.

Meredith's lab maintained "tracking sheets" for monitoring work from day to day. The tracking sheets record "which libraries we're making, who did it, when they started, how much they started with." As well as containing handwritten notes on a worker's activities, the tracking sheets interface with the barcode system: in order to use a reagent, the worker must peel off its barcode and attach it to the tracking sheet; at the end of the week, the tracking sheets are scanned and the inventories of reagents updated. The electronic record of the tracking sheet is then linked to electronic files containing pictures of gels and results of QC tests. This database is maintained in the business operations software SAP. Without such a sophisticated record, Meredith tells me, high throughput would be impossible: the database allows "fairies" (who resupply the lab) to make sure the lab never runs out of anything. "We don't stop for anything," Meredith reassures me.

> Before when I started here there was no standard tracking sheet. People would do your very common diary-type that molecular biologists do in the lab, page numbers . . . and they just say, "this is what they did today." . . . Which is great, except when you need to troubleshoot and figure out why this is so good or why this is so bad, you go back, and you need to go back many pages, and many times people didn't think that was a very important piece of information to keep. . . . There is not much reliability in

the data. . . . When you do a standard tracking sheet, you know it's there, and it's always there. You also enforce, or at least you can see, that it's been done the same way over and over again. This is a production environment and for us variability is hard to deal with, we want to have as little variability as possible and standard tracking sheets are very good for that.

The detail with which such tracking is performed is illustrated in the kinds of checklists that Vokoun proposed for the MBPG. The checklist in figure 3.7 shows that workers had to account for the numbers of tips, wipes, tubes, and caps at their workstation each day. Their managers then used a sheet to score their work on the basis of "shiny clean" floors, "unused pipettes, tools, [or] fixtures" cluttering their workspace, maintenance of checklists, and so on.[39]

All this monitoring depends critically on machines. It is computers that maintain not only the detailed monitoring data, but also the careful control over space and people. "Our life is spreadsheets," Meredith told me simply. "We love spreadsheets, we hate spreadsheets." But Meredith also told me that some of their needs for managing data had far outgrown the capacity of spreadsheets. By now it was really databases that ran the lab: SAP and the Broad's laboratory information management system (LIMS), called SQUID. In one way at least, the cephalopod name is appropriate: SQUID's tentacles extend in all directions into all corners of the laboratory, sucking data back to a central repository. Any sample that passes through the lab leaves its trace in SQUID—the history of its movement is recorded in painstaking detail. Natalie, an associate director of the Genome Sequencing Platform, described her work coordinating and managing sequencing projects. Projects were initiated by generating records in SQUID, projects were monitored by watching their progress through SQUID on a daily and weekly basis, and projects ended when they were removed from the database. At the Broad, the production of sequence becomes an information management problem. The ability to manage leanly, to create spaces and people amenable to the principles of operational analysis, means having the ability to measure, to quantify, and to track.

The Broad's raw material is samples (it deals with thousands); its products are bases of DNA sequence (it produces billions per year); in the middle, petabytes of data are generated. Measuring, quantifying, and tracking are possible only with computers. It is computers, particularly large and sophisticated databases, that have allowed the techniques of production management to be imported into molecular

5S Master Checklist - MBPG

≣BROAD INSTITUTE

November 2004

Section: MBPG	Today's Score:	65
Scored By: M. Vokoun		
	Previous Score:	n/a
Date: 12/1/04		

5S	#	Check Item	Description	0	1	2	3	4	Section Total
Sei (Sort)	1	Unneeded material, low quality or broken parts/WIP	Does the inventory or in-process inventory include any unneeded material or broken parts?			■			
	2	Machines or other equipment	Are there any unused machines or other equipment around?			■			
	3	Pipettes, tools or fixtures	Are there unused pipettes, tools, fixtures or similar items around?			■			12
	4	Unneeded items	It is obvious which items have been "red tagged" as unnecessary?			■			
	5	Instruction signs or other paperwork	Has establishing 5S left behind any useless instruction signs or other paperwork?				■		
Seiton (Straighten)	6	Location/item labels	Are shelves and other storage areas labeled with location/item indicators and addresses?				■		
	7	Notebook/paperwork areas	Are lab notebooks, filing cabinets, or the paperwork areas organized/labeled?			■			
	8	Quantity indicators	Are the target and/or max-min quantities indicated on all consumable materials?			■			13
	9	Walkways, storage areas, and in-process inventory areas	Are lines or other markers used to clearly indicate walkways and storage areas?				■		
	10	Equipment, pipette, and lab coat arrangement / organization	Are equipment, pipettes, & lab coats placed back into their designated locations?				■		
Seiso (Shine)	11	Waste, water or tips on floors	Are floors kept "shiny clean"?				■		
	12	Bench / equipment cleanliness	Are lab benches and equipment wiped clean?				■		
	13	Equipment inspection/maintenance	Do employees clean their equipment while checking them?			■			13
	14	Cleaning tasks	Instead of cleaning up messes, have people found ways to avoid making messes?			■			
	15	Wires / power cords / cables	Are wires, power cords, and cables secured or tie wrapped?				■		
Seiketsu (Standardize)	16	Standards / protocols	Are all standards and protocols known and visible?				■		
	17	Roles posted	Have roles/responsibilities been assigned and posted to maintain the first 3 S's?				■		
	18	Check lists	Is there a check list where appropriate?					■	14
	19	Information	Is all necessary information (signage, instructions, safety, etc.) posted?				■		
	20	Work environment	Is ventilation and lighting adequate for the work?				■		
Shitsuke (Sustain)	21	Cleanliness	Do employees habitually clean lab benches, wipe equipment, and sweep the floors without being told?				■		
	22	Red tag procedure	Are red tag procedures followed?			■			
	23	Time keeping	Do people keep their appointments/meetings and take their breaks on time?				■		13
	24	Awareness of standards / protocols	Are new / existing people aware of the standards and protocols?				■		
	25	Following protocols & standards	Do people follow established protocols & standards?				■		

| Matthew R. Vokoun ▓ 11/3/04 | Overall Total | 65 |

FIGURE 3.7 The Molecular Biology Production Group's 5S master checklist. Workers are scored on such questions as, "Are there unused pipettes, tools, fixtures, or similar items around?" or "Are lab benches and equipment wiped clean?" or "Are all standards and protocols known and visible?" (Vokoun, "Operations capability improvement." Copyright Massachusetts Institute of Technology. Reproduced with permission.)

biology. Consonant with the spirit of lean management, the central role of the computer does not mean that workers have lost all agency in the sequencing process; indeed, individual workers have a high level of responsibility for monitoring and improving their own work. It is the computer, however, that draws all these individual actions together into a collective whole—it is the computerized monitoring and management that integrates a person's work into that of the team. In other words, it is through the computer that individual labor becomes "productive."

What are the consequences of this borrowing from business? What difference does it make that the Broad is organized more like Toyota than the Pasteur Institute? First, it has changed the notion of valuable work in biology. The work accorded value at the Broad Sequencing Center is not the highly individualistic, highly innovative work of the traditional bench scientist. What is valued instead is teamwork, attention to detail, precision, and efficiency. Second, work at the Broad is based on a new accounting of biological work: the lab is funded not according to how many papers it publishes or how promising its research seems, but on the basis of dollars per base. The National Human Genome Research Institute (NHGRI, from which a large proportion of the money comes) favors the Broad because it can offer its product (sequence) at a cheaper rate than its competitors.[40] This accountability is passed down through the hierarchy of the organization to the bench worker, who must also be held to account for his or her own productions.

Third, what constitutes "progress" in advancing biological knowledge has changed: progress can be understood as the accumulation of more and more sequencing data at an ever-decreasing cost. The immediate goals of day-to-day work are discoveries that will increase output by decreasing variability, rather than making fundamental breakthroughs or shifting work in a qualitatively new direction. Fourth, the culture of the Broad Sequencing Center suggests a shift in what sorts of people are doing biological work as well as changes in the distribution of labor. Whereas before, biology was performed almost exclusively by PhD scientists (and their graduate students and postdocs), biology at the Broad demands a workforce that is not trained only in biology and whose skills might be transferable to a range of industries. While these workers are busy with the laboratory/manufacturing work of sequencing, the PhD biologists are engaged in a distinct set of tasks, often physically and intellectually removed from the lab bench. Finally, this new kind of biological lab has become a space of surveillance to an extent previously unusual in the sciences. Through information, people, objects, and spaces are constantly monitored; every pipette tip wasted or moment spent chatting with a colleague leaves a discernible informatic trace. Everything must be accounted for. Here, biology has become a sort of informatic Panopticon; Natalie told me that she liked her job because the 30,000-foot view of the Broad's work provided an appealing sense of control. Doing good and interesting work means keeping watch and being watched.

All these observations suggest that biological knowledge production—in genomics at least—has undergone a fundamental transformation. Authorized and valuable knowledge is high-quality, high-quantity knowledge; it must be checked for errors, scrutinized, and monitored throughout its production. It must be accountable, both in the sense that it be carefully costed and that its provenance (recorded in the database) be rigorously checked. It was the computer—as a data management machine—that allowed the concepts of lean management, Six Sigma, and so on to be implemented in biology; the mass production of (sequence) data as a product required computers in order to make the sequencing process visible, manageable, and accountable. At the Broad, the organization of biology in accord with "business principles" has deeply shaped practices of knowledge making. Computers have created new ways of making authorized and valuable knowledge through careful accounting and management of data.

Conclusions

Wet-lab biologists, computational biologists, and system administrators use and relate to the spaces around them differently. These different workers are acutely aware of these differences—the status that is attributed to biological knowledge is dependent on the spaces from which it emerges. The highest status is accorded to knowledge produced in the highly visible front spaces and wet labs, while lower status accrues to knowledge associated with the sorts of technical production that take place in the back spaces of the Broad. Moreover, the layout and design of the spaces recapitulate and reiterate the kinds of knowledge that are produced within them: back spaces promise reliable practices of replication and production, while front spaces adhere to more stereotypical images of scientific practice. And the movement of people and data through and between spaces plays a crucial role in validating and certifying knowledge.

All economies benefit some and disadvantage others: they are sets of power relations. The production and consumption of bioinformatic knowledge confers prestige on academically trained biologists while restricting credit for data producers. As a consequence, bioinformatic work continues to be described and understood as fundamentally biological—that is, differences in visibility and power cause us to underestimate and undervalue the contributions of managers, physicists, engineers, technicians, mathematicians, and computer scientists to this new kind of biology. These differences have significance for those who might point to the need for greater value and emphasis to be placed on new skills of data management and data curation in biology. The argument here suggests that there is a reason why such pursuits remain undervalued: namely, that the status of biological knowledge as biological knowledge partly depends on the hierarchy between producers and consumers.

But the biological work of the Broad consists equally of what is going on in the back spaces. Following the data into spaces such as the Broad Sequencing Center reveals the emergence of new forms of practice with new forms of value that are a critical part of bioinformatic biology. The emergence of these practices and values is a result of computers, which have transformed "production" from a "bad" to a "good." The organization of space and work has been transformed by the power of computers to track, organize, sort, and store.

The mass and speed of bioinformatics that we encountered in chapter 2 does not just apply to the accumulation of data about DNA sequences. Computers are also able to collect and manipulate all kinds of other data: about samples, workers, reagents, error rates, and so on. These data are also contributing to the transformation of biological work in terms of volume, speed, and efficiency. The history of the computer suggests the reasons for this: from the 1950s onward, computational practices were linked to and evolved from efforts to rationalize government bureaucracy and commerce. For instance, the Treasury of the United Kingdom used punched-card-based computers for accounting, payroll, and the production of statistics.[41] And it has been argued that in the 1950s, "the computer was reconstructed—mainly by computer manufacturers and business users—to be an electronic data-processing machine rather than a mathematical instrument."[42] Many of these "calculators" were used at first for accounting purposes and later for administration and management. Computers were used for payroll calculations, sales statistics, and inventory control.[43] The advent of the UNIVAC and the IBM 701 in the early 1950s made computers valuable to business as machines able to "automate" and speed up routine tasks. At General Electric (where one of the first UNIVACs was installed), the digital computer was used first to replace salaried clerks (and their overhead costs) and later for long-range planning, market forecasting, and revamping production processes.[44] IBM—and other companies attempting to compete with it—had to design the computer for the needs of business: alphanumeric processing, checking and redundancy mechanisms, "buffers" for high-speed transfers, magnetic tape storage, and records of variable length. Such features later became commonplace in all digital computers. As James Cortada has argued, "a quick look at how computers were used suggests that the history of the digital computer is every bit as much a business story as it is a tale of technological evolution."[45] In the 1950s and 1960s, the computer was developed as a tool for exercising close control over a corporation and making business more efficient. This preoccupation was reflected in its design.

This history has conditioned the role that computers have come to play in biology—that is, as a tool for speed and efficiency. For example, attempting to search biological databases by hand and eye was possible in principle, but impossible in practice; computer techniques brought speed, efficiency, and accuracy to the process. As such, the principles and practices of bioinformatics were always and already "industrial" in an important sense: they were attempts to streamline information flow and knowledge production in biology. Computers are tools of business, and they demand and enforce the kinds of practices that have transformed biological work over the last two decades, reorienting it toward speed, volume, productivity, accounting, and efficiency. As this chapter has shown, however, it is not only the samples and sequences, but also the laboratory itself and its workers, that are managed and accounted as data. The computer brings changes that are at once social, technical, spatial, and epistemic. Bioinformatics entails reorganizations of space and people in tandem with reorganizations of practice and knowledge.

These findings obviously rest heavily on observations of the Broad Institute, but similar patterns can be discerned elsewhere. The Sanger Institute (Hinxton, UK), the J. Craig Venter Institute (Rockville, MD), the Genome Institute at Washington University (St. Louis, MO), the Joint Genome Institute (Walnut Creek, CA), and the Beijing Genomics Institute (Shenzhen, China) all have large-scale production sequencing facilities.[46] The aim of each is to produce high-volume and high-quality product at low cost.

Rather than a biology oriented around individual investigators and intradisciplinary work, bioinformatics is a biology performed by large, multidisciplinary teams, oriented toward efficiency rather then reproducibility, measured by accounting and QC rather than by peer review, and ordered by information systems. Toyota's concept of "lean production," as deployed at the Broad Institute, suggests a label for the ways in which the making of "biovalue" depends on new forms of practice as well as new regimes of circulation. A "lean biology"—a biology stripped back to its essentials, reduced to its elements (in particular, data and sequence), made efficient—is the kind of biology required to make bioinformatic knowledge. "Lean biology" is a mode of practice that is designed to harness the productivity of samples, workers, and computers—to maximize the production of data and knowledge by ignoring or stripping or abstracting away all functions extraneous to the generation of these products.[47] In doing so, it creates new kinds of biological knowledge that are rooted in the practices of industrial management and in informatics. Lean biology—the sets of practices that produce bioinformatic knowledge—is a way not only of describing this trend, but also of suggesting how it is coupled to the commoditization of life in the age of biotechnology and biocapital.[48]

4   Following Data

Telling stories about how people move around inside the physical spaces of laboratories and sequencing centers reveals only some of the topology of bioinformatics. Twenty-first-century science involves a further set of spaces. In bioinformatics, the relationships between spaces are not only physical, but are also influenced by other kinds of proximities: Ethernet cables, microwave links, and shared computer networks form a second kind of "space" that must be mapped and taken into account. This chapter will pay close attention to these virtual spaces—how they are arranged, who can access them, and how people interact within them.[1] The landscape of scientific communication and the possibilities for collaboration, interaction, and exchange are fundamentally reworked by electronic networks; these spaces must be considered autonomous means through which scientific actors and objects can be ordered, distributed, and brought into proximity with one another. In other words, they present new modes through which scientific knowledge can be produced and certified.

As one might expect of a digital discipline, much of the important space and motion in bioinformatics is virtual: shared disk drives, local networks, server farms, compute farms, and the World Wide Web constitute another, less immediately apparent, landscape in which bioinformatics is enacted. In the labs I visited, these virtual spaces were not only harder to find and harder to access, but more intricate and more difficult to navigate, than the physical spaces. Indeed, a large part of learning to do bioinformatics is the process of learning to find one's way around the appropriate virtual spaces—having the expertise to know *where* some piece of information or code might be was often as valuable and as valued as knowing what to do with it. This chapter draws on my experiences at the Broad Institute and the European Bioinformatics Institute (EBI) in order to describe what these virtual spaces of bioinformatics look like, how people move around in them, and how this movement constitutes the *doing* of bioinformatics.

The first part of this chapter details how biology becomes virtual in the first place—how material samples become data as they pass through the sequencing "pipeline." This process is at the center of bioinformatics: showing how the material becomes the virtual is to show how biology becomes digital. What becomes apparent in the description of the pipeline, however, is that this flattening of objects into data is a complex and contested process—samples do not automatically render themselves into zeroes and ones. Although a pipeline suggests linearity, what we find here is that the process is "out of sequence"—it is messy and contingent. Following the data to their origins shows how much their generation depends on care and judgment and how much work must be done to make standardized, virtual biological objects that are amenable to computers and networks.

In the second part, I examine another way in which biology is rendered into data. Producing "ontologies" means creating standardized common languages for speaking about biology. The uniformity and universality of biological objects is manufactured during bioinformatic work. Such standard, computable objects are necessary for bioinformatic work to be possible. Bioinformatics standardizes and flattens biological language into data. Third, I examine how bioinformatic objects are moved around. The rendering of biology into standard, computable formats allows knowledge to be produced by the careful movement and arrangement of data in virtual space.

What emerges here is a new picture of biological work and biological knowledge. Biological work consists in the rendering of biological stuff into data and the proper arrangement of these data in virtual space. Neither the creation of data nor their movement in virtual space is a wholly linear, fluid, or frictionless process. Data must be carefully formed into their digital shapes, and their arrangement must accord with standardized structures such as standardized data formats, ontolo-

gies, and databases. These structures impose constraints on what biological objects are and how they can be talked about.

Pipelines

At the center of bioinformatics is a process that transforms the real into
the virtual—that renders biological samples into data. The "pipeline" is
a spatial metaphor for a process that recurs often in bioinformatics. As
such, it is also an appropriate metaphor to use for describing the transition from old to new forms of biological work. The pipeline moves us
from the material to the virtual and from the lab bench to the computer
network.

In biology, "pipeline" is a word used to describe the series of processes applied to an object in order to render it into some appropriate final form. Pipelines can be either physical (involving transmission
and transformation of actual objects) or virtual (involving transmission
and transformation of data), but they are often both, and they most
often describe the processes through which actual objects (DNA from
an organism) are rendered into virtual form (DNA in a database). The
pipeline is the method through which the actual becomes the virtual,
through a carefully ordered set of movements through physical and virtual space.

When we think of a pipeline, we might immediately think of an oil
pipeline or a water pipeline, in which a liquid is transported directly
along its length. Readers familiar with computers might also think of
the ubiquitous "pipe" command in Unix (represented by "|") by which
one program or command takes as its input the output from the previous program; by "piping" several programs together, the user creates a
space through which data can flow. In both these cases, there are two
key concepts. First, pipelines are directional: the water or the data must
flow down the pipe, following the single path laid out for them and
never moving side to side or going backward. Second, pipelines require
liquidity: pipes are generally inappropriate for moving solid objects, and
piping programs together requires that inputs and outputs of adjacent
programs in the pipe be designed to match one another.

In 2008, I spent several months observing the "sequencing pipeline"
in action at the Broad Institute. At the Broad Sequencing Center, almost
all the work is based around the pipeline—it is the metaphorical object around which work is ordered. The pipeline is the set of methods
and processes that are used to transform molecular DNA into sequence
data in appropriate forms for submission to an online database such as

GenBank. Samples arrive at the sequencing center in many forms: cells growing in media, microorganisms, blood, tissues, or even whole large organisms.

It is the task of the Molecular Biology Production Group (MBPG) to extract the DNA and prepare it for sequencing. The MBPG's role consists of three stages. First, the DNA preparation team subjects the incoming DNA samples to a number of tests to check their purity and quality. They then shear the DNA into random pieces of a size suitable for sequencing, using a purpose-built machine that subjects the DNA to hydrodynamic forces. The DNA is then stored in a −20°C freezer in the MBPG lab area. Second, the ligation team is responsible for the 5–6-day-long process of library production. A library of DNA is a large collection of short DNA fragments that together represent a complete segment of DNA (a whole chromosome, for example); since a library is constructed using many copies of the same segment, most of its parts will be represented in many of the small pieces. The DNA pieces are chemically treated so that they are incorporated into other specially engineered, ring-shaped pieces of DNA called plasmids—this process of fusing DNA is called ligation. Completed libraries are also stored in freezers. Finally, the transformation team produces finished agar plates for handoff to the Core Sequencing Group. *E. coli* bacteria are mixed into the plasmid solution and, by rapid heating or electric shock, induced to take up the plasmids (transformed). Workers must then spread the bacteria-containing solution thinly over a 9-by-9-inch agar plate infused with an antibiotic. The engineered plasmid includes an antibiotic resistance gene, so *E. coli* that have not taken up a plasmid will die, while those that have incorporated a plasmid will grow into colonies as they are incubated overnight. Each resulting colony on the agar plate should contain many copies of a single DNA fragment in its plasmids.[2]

Each of these tasks is delegated to a team consisting of four or five workers, usually with one acting as a coordinator. Every sample receives a barcode when it enters the pipeline. As a sample moves around the laboratory, this barcode is repeatedly scanned, and the results are stored in a laboratory information management system (LIMS). For example, a sample is scanned when it is put into or taken out of a freezer and when it undergoes a particular process, such as ligation. Through the LIMS, it is possible to track the status and location of any sample in the sequencing center at any time. By querying the LIMS, workers can discover what work is available for them to do and where to find it. For instance, a member of the ligation team can find the freezer locations of all the samples that have undergone DNA preparation and are ready for ligation.

In these early steps, much attention is paid to space and motion. The time and effort required to move samples through the lab is carefully monitored and analyzed. The locations of equipment (centrifuges, freezers, etc.) and lab benches are optimized so that workers can move the samples quickly and efficiently through the lab spaces. These processes are constantly scrutinized in order to maintain the highest levels of quality and efficiency. Incubation times, solution concentrations, and other details are monitored and modulated to increase yield.

Once the agar plates are incubated, the responsibility of the MBPG ends and the Core Sequencing Group takes over. Its work can be divided into six steps. First, in the "picking" step, a large robot designed by a company called Genetix uses a high-resolution camera to scan the agar plates to produce digital images. These images are processed by software that identifies colonies that are good candidates for picking—that is, colonies that show good growth and are not too close to one another. A robot arm then moves a 96-pin head over the plate, picking samples from the selected colonies and transferring each to an individual well of a 384-well tray or plate containing glycerol solution.[3] The agar plates are then discarded, and the 384-well plates are incubated for a further 15–18 hours. The samples are then once again placed in a freezer. Second, the "TempliPhi" step generates further amplification, using a PCR-like process to increase the amount of DNA about 10 millionfold in 4 hours. Third, in the "sequencing" step, specially designed nucleotides tagged with fluorescent dyes are added to the samples and are incorporated into the DNA molecules. These incorporations are the sequencing reactions themselves, the results of which are detected in step 6. In order to get a sense of what is involved in this process, I quote from a detailed description:

> The sequencing team starts by transferring the DNA from the TempliPhi Eppendorf Twintec™ plate into two other Eppendorf Twintec™ plates and diluting the DNA mixture with water. The top of the red plate (the TempliPhi plate) is transferred into the top and bottom halves of one blue plate, so there are two copies of DNA in one blue plate. The bottom half of the red plate is transferred into another blue plate, and the red plate is then discarded. Once this is done, two types of dye are added to each blue plate; "forward" sequencing dye is added to the top half of

the plate, and "reverse" sequencing dye is added to the bottom half of the plate.[4]

The fourth and fifth steps are "ethanol precipitation" and "elution." respectively. Both of these chemical processes are designed to remove excess dye and other contaminants. All five of these steps are far from automated: each involves careful laboratory work, testing, and judgment calls. In picking, the agar plates are "visually checked" for approximate accuracy of counts and spacing; the 384-well plates are checked for a "hazy" appearance (indicating a growing colony) before sequencing; solutions used in the TempliPhi steps are checked each morning (randomly selected plates are weighed to ensure that correct amounts of reagent are being added).[5] This is not a straightforward sequence of steps, but rather an interlocking set of checks and tests that require knowledge of the process chemistry.

It is in the sixth step, "detection," that the samples disappear wholly into the virtual. And it is only in this step that the samples are introduced into the sequencing machines. About 120 Applied Biosystems 3730 detectors, each costing hundreds of thousands of dollars, sit together in a large single room—each is about 6 feet high and has a footprint of roughly 3 feet by 2 feet. Because of the cost of the machines and the speed at which they process samples, detection is both the rate-limiting and the cost-limiting step. As such, the detectors are operated twenty-four hours per day, seven days per week. Inside the machines, the dyes attached in step three are excited by a laser and detected by a CCD camera. In typical detector output, the four possible nucleotide bases are represented by lines of four different colors on a chart (red, green, blue, black).

These colored graphs are known as the "raw sequence traces." They are stored by the sequencing machines, but they are not yet "finished" sequence. As the raw sequence traces are generated by the detectors, two computational steps are automatically performed. The first is known as "base calling" and is usually done using software called "Phred." Phred analyzes the raw sequence trace images and, based on the height and position of each peak, makes a "call" as to what the correct base is for each position along the sequence. In other words, it converts an image file (the four-colored chart) into a text string consisting of the familiar As, Gs, Ts, and Cs. It also assigns a quality score to each base, reflecting its confidence in the call that it makes. The second automated step is "sequence assembly." Recall that in the very first stages of the sequencing process, before the DNA was spliced into plasmids, the entire sequence was broken into random fragments. In order to generate the whole sequence, the individual reads (one from each well of a tray or plate) must be reassembled. A program called "Phrap" is used to do this. Phrap uses the base calls and the quality scores from Phred to determine the most likely assembly by searching for overlapping segments (it takes into account the fact that in regions where quality scores are low, overlaps could be due to random errors). Both Phred and Phrap are open source software packages developed in the early 1990s by Phil Green and his colleagues at Washington University, St. Louis.[6]

Despite the sophistication of these algorithms, they are usually unable to match up every sequence—for almost every genome, gaps remain. It is the job of the "finishing" team to take over where Phrap leaves off and patch up the gaps. There are about forty finishers at the Broad; their work is highly specialized and requires detailed knowledge of the biology of the organisms being sequenced, the intricacies of the sequencing process, and the assembly software. By querying the LIMS, a finisher can find samples in the queue that require finishing work. During my fieldwork at the Broad, I spent several hours watching a highly experienced finisher at work. After retrieving the relevant sequences from the LIMS database, the finisher imported them into a graphical user tool that shows the overlapping sequence regions one above the other. Scrolling rapidly along the sequence on the screen, the finisher made quick decisions about the appropriate choice of base at each position where there seemed to be a discrepancy. His experience allowed him to tell "just by eyeballing" where Phrap had made a mistake—by placing TT in place of T, for instance. The graphical tool also allowed the finisher to pull up the raw sequence traces—which was sometimes necessary in order to make a decision about which base call was the correct one. Where gaps existed, the finisher imported sequence data from sources that had not been used in the assembly (an online database, for example) in order to begin to fill the hole. This work is often painstaking and relies crucially on the finisher's judgment. The following quotation describes the finisher's reasoning process as he changed a single G to a C in the sequence:

> Here now is a discrepancy in the consensus [sequence]. Usually it's where we have a whole string of Gs like we have here; these reads here are fairly far away from where the insert is—I can tell just by their orientation—they've been running through the capillary [the key component of the detector] for a while when it hits one of these stretches . . . so it's basically calling an extra

G even though these high-quality reads—you can tell by the color—are clearly correct, clearly calling a discrete number of Gs, and so again this one is a C.

Where it seems that there are insufficient data to fill a gap, the finisher may request further laboratory work on that sequence region—such requests are passed through the LIMS. Making these requests also requires careful consideration:

> Finishers often find themselves in a catch-22: to select an appropriate laboratory procedure, they must understand the underlying sequence, but the sequence is missing. In practice, they must make educated guesses about the underlying sequence and the likelihood that various laboratory techniques will succeed. Their decision is influenced by the condition of the DNA near the gap; it is also influenced by the ability of their informatics tools to highlight those conditions. Most importantly, finishers' decisions are guided by their skill and experience: whereas some experienced finishers may be able to close a gap based on the information already present in an assembly, less experienced finishers may feel they need laboratory work.[7]

Under its contracts with the NHGRI and other funding bodies, the Broad's standard for "finished" sequence requires no more than one error in every 10,000 bases. A recent audit found error levels around one in every 250,000 bases. Once a finisher is satisfied with the piece of sequence on which he or she is working, it is resubmitted to the LIMS. From there it must pass through a "final review" by a different finisher before it can be submitted to GenBank as a "finished" sequence.[8] During this review the finisher runs a number of scripts on the sequence to check for quality and consistency and to align the piece of sequence with the entire genomic build. The submission process is also automated: a sequence is passed through a script that automatically generates the required metadata, such as the coordinate systems, the sequence length, and the author names. This script also does a final check for gaps and writes standard features into a GenBank submission file. All progress, including the GenBank submission number, is recorded in the LIMS. Finished submission files are automatically uploaded to GenBank via FTP at 11:00 p.m. each day, where they are available to the world at 8:00 a.m. the following morning.

The integrity and fluidity of the sequencing pipeline are maintained by the LIMS. Without this computer tracking system, the various activities of the pipeline would remain disconnected from one another. It is the LIMS that makes it possible to think of a discrete-sample-thing flowing through the pipeline; it is only in the computer that the place-to-place movement of the samples is made sense of, or makes sense. This makes it problematic to think of the process as a clear-cut transition from actual sample to virtual sequence inside the detector; as soon as a sample is barcoded and entered into the LIMS, it becomes a virtual as well as an actual object. The actual object in the lab would be meaningless without its virtual counterpart linking it to the pipeline. As the actual-virtual object progresses down the pipeline, however, it does become increasingly liquid—as it is further disconnected from its actual presence, it is transformed into a more mobile form. Images such as figure 3.3 suggest the intricate labor that must be performed in the early stages of the pipeline as the sample is tracked in its minute movements around the lab. By the finishing stage, however, the sample can be effortlessly pulled up onto a screen and, finally, dispatched instantly across hundreds of miles to the GenBank database in Bethesda, Maryland. The sequencing pipeline can be understood as a process of enabling DNA to travel.

But the pipeline does more than make things flow. The metaphor also serves to obscure the transformational and judgmental aspects of the sequencing process. Real pipelines merely transport a uniform substance from one place to another. The metaphor of a sequencing pipeline (rather than a sequencing production line, for instance) suggests that the extracted text was already inside the body of the sequenced subject, waiting to be piped out (like oil or natural gas); it suggests that the As, Gs, Ts, and Cs are merely transported from the organismic body to hard drives in Bethesda. The pipeline metaphor allows biologists to understand sequencing as travel through space, rather than as an active process of extraction and construction shot through with difficult manual tasks and active judgment calls. Those directly involved in the day-to-day work of the pipeline rarely experience it as a linear flow. Figure 4.1 shows a process flow diagram depicting a far more intricate set of actions and interconnections than is suggested by a pipe. Organizational and workflow charts on the walls of the Broad depicted something more like a dense network than a pipeline.

The metaphor, however, serves to conveniently collapse the sequencing process for those who use sequence data; imagining the sequencing process as a linear flow serves to de-problematize the actual-to-virtual transformation and allows the sense that sequencing is an automatic, black-boxable activity. Such an elision is crucial in bioinformatics, for

Ligation Process Flow Diagram – Level 3

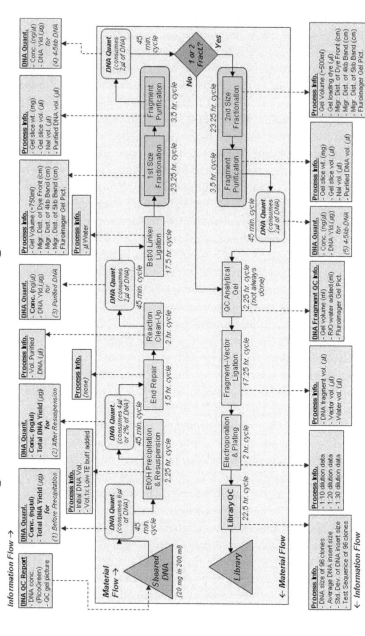

FIGURE 4.1 Ligation process flow diagram. The ligation process is only a small part of the Broad's sequencing "pipeline." Note that the details of the process are not strictly linear (it branches, and some steps are marked "not always done"). Note too that the process is simultaneously material and informatic. (Vokoun, "Operations capability improvement." Copyright Massachusetts Institute of Technology. Reproduced with permission.)

which "the sequence" is the central object. Understanding sequencing as a directed, unproblematic process—a simple linear motion through space—allows the output products (the digital renderings of the sequences themselves) to be plausibly used as proxies for the input materials (wet biological samples). Almost all of bioinformatics depends on this fact that sequences can be relied on as straightforward extractions of biological material into biological data. This section's detailed description of the sequencing pipeline shows, however, that the production of biological knowledge (in this case, sequence) depends crucially on carefully contrived movement through space. Such movement is far from automatic or linear—it is shot through with judgment calls and contingent processes. The production of "fluid" and "universal" bioinformatic objects depends on a highly situated process through which this "fluidity" and "universality" is constructed; bioinformation can travel and flow only because its movement through the pipeline serves to obscure its solidity and situatedness.

The pipeline metaphor is not intended to completely capture the features of a biological specimen. Indeed, such a task would be impossible. Rather, the actual-virtual transition reduces the sample to a digital trace, flattens it into a text that can be computed, communicated, and manipulated in virtual space. Sequence data, consisting of a base call and a quality score, already constitute a structure to which biological sequences must conform. Following the data to their place of production shows how they are constructed for a particular purpose; they are crafted as data objects ready to flow through computers and networks.

Standards and Ontologies

Simply producing sequence data and putting them in GenBank would not be useful for biological research unless bioinformaticians and biologists already agreed about standards and formats for writing, storing, and reading such data. GenBank is doing far more than just storing biological data—it is enacting common standards for the reading, writing, and storage of biological information. This section will examine how bioinformaticians go about developing and enforcing common languages. Such languages are required not only for communication, but also for making knowledge. When language is flattened into data, it becomes possible to create ways of agreeing about objects and their proper descriptions.

For nucleic acid sequences themselves, a standard shorthand was already in place: A, G, T, C, and U were in use before large-scale sequencing began. As DNA sequencing became ubiquitous, the genetic code was supplemented by other letters, such as R (G or A), B (G, T, or C) or N (A, G, T, or C). Such tolerance for ambiguity reflected the fact that sequencing was still sufficiently expensive that some information was better than none. Such a code is already a significant abstraction from a "physical" DNA molecule, which might contain nonstandard nucleotides, epigenetic markers, three-dimensional conformations, or other chemical variations.[9]

Despite the standard AGTC shorthand, variation in coding was still problematic for bioinformaticians. Different software programs, for example, required sequences to be submitted in different formats. By the late 1980s, however, the superior speed of Bill Pearson and David Lipman's FASTA program meant that its input format gradually became a standard for the encoding of sequences. Even though FASTA was superseded by BLAST, the FASTA format persisted.[10] The format consists of a first line that begins with a ">" on which information identifying the sequence is listed, followed by the sequence itself on subsequent lines. The appearance of another ">" in the file indicates the beginning of a new sequence. Although FASTA has been criticized because the identification line lacks a detailed structure, it remains the de facto standard for sharing and transferring sequence data. Its simplicity makes it particularly attractive to programmers who want to be able to parse sequence data into their programs quickly and easily. It is a simple but powerful data structure to which sequences must conform.

The ready existence of a widely agreed-upon code has made the sharing of sequence data relatively straightforward. The sharing of other kinds of biological data, however, has required more elaborate schemes. In particular, it is what are known as "annotation data" that have caused the greatest problems. Annotation data include all the information attached to a sequence that is used to describe it: its name, its function, its origin, publication information, the genes or coding regions it contains, exons and introns, transcription start sites, promoter regions, and so on. The problem is that such data can be stored in a variety of ways; different descriptions in natural language (for example: "Homo sapiens," "H. sapiens," "homo sapiens," "Homo_sapiens"), different coordinate systems, and different definitions of features (for instance, how one defines and delimits a gene) inhibit compatibility and interoperability. There are two kinds of solutions to these problems, one of which I will call "centralized" and the other "democratic."

The democratic approach is known as a distributed annotation system (DAS). As it was described in a seminar at the EBI, a DAS is like a Web 2.0 "mashup," like using Google to create a map of local pubs by pulling in data from different sources. The idea is that the data remain in various "native" formats in a large number of geographically dispersed repositories; the DAS server, however, knows how to talk to each repository and extract data from it on the fly. The EBI's main database, known as Ensembl, works largely through a DAS. When a feature is requested through the Ensembl website, the website creates a standard URL that tells the DAS which information to retrieve; the DAS then queries the remote database and sends back the relevant data in XML format, which can be interpreted and displayed by the Ensembl website. The Ensembl DAS does not aim to translate all the data into a common language or format—instead, it can display data from multiple sources side by side on a website. This method not only sidesteps problems of translation, but also avoids (since data are requested in real time over the web) time-consuming efforts to keep data up to date. The inventors of the Ensembl DAS system were explicit with their intentions for the kinds of biological work they hoped it would promote:

> DAS distributes data sources across the Internet improving scalability over monolithic systems. This distribution of data encourages a divide-and-conquer approach to annotation, where experts provide and maintain their own annotations. It also permits annotation providers to disagree about a particular region, encouraging informative dissension and dialogue. The separation of sequence and map information from annotation allows them to be stored and represented in a variety of database schema. A number of different database backend alternatives could arise. The use of links as a method of referencing back to the data provider's web pages provides even greater power of expression and content control.[11]

The DAS is not only a technical solution but also a mode of political organization for biology. The creators of the DAS imagined a democratic biology in which the task of data management is shared and knowledge is a product of debate and negotiation. The DAS does not allow biological data to travel anywhere—local databases must be compatible with the XML standards that the DAS server uses. However, the DAS attempts to be minimal in its approach, requiring a relatively small investment on the part of the local database. The DAS represents a simply constructed network that is built using local materials and using local methods; it is highly heterogeneous but can include a large amount of data because it does not need to be rigorously maintained or policed.

Centralized solutions, on the other hand, aim to bring all the data into a common format at a single location—software is written to translate data from various formats and deposit and maintain them in a master database. These are the most ubiquitous and fastest-growing solutions to the problems of sharing biological data. These centralized solutions, called "ontologies," manifest a philosophy opposite to that of a DAS, as they tend to be top-down and heavily managed. Ontology is the branch of philosophy that deals with what things exist in the world: it is "the science of what is, of the kinds, structures of objects, properties, events, processes, and relations in every area of reality. . . . Ontology seeks to provide a definitive and exhaustive classification of entities in all spheres of being."[12] For computer scientists, predominantly those concerned with information management and artificial intelligence, "ontology" has a slightly different meaning:

> a dictionary of terms formulated in a canonical syntax and with commonly accepted definitions designed to yield a lexical or taxonomical framework for knowledge-representation which can be shared by different information systems communities. More ambitiously, an ontology is a formal theory within which not only definitions but also a supporting framework of axioms is included.[13]

Why would computer scientists care about making such dictionaries? If computers are to be able to reason with information, the language used to communicate such information to the machines must be standardized. Several ambitious attempts have been made to build exhaustive vocabularies for large-scale business enterprises. In 1981, the firm Ontek began developing "white collar robots" that would be able to reason in fields such as aerospace and defense:

> A team of philosophers (including David W. Smith and Peter Simons) collaborated with software engineers in constructing the system PACIS (for Platform for the Automated Construction of Intelligent Systems), which is designed to implement a comprehensive theory of entities, ranging from the very concrete (aircraft, their structures, and the processes involved in designing and developing them) to the somewhat abstract

(business processes and organizations, their structures, and the strategies involved in creating them) to the exceedingly abstract formal structures which bring all of these diverse components together.[14]

This was an attempt to create a "theory of the world" that could be used by computers. In biology, the problem is not only getting computers to work with biological data, but also getting biologists to work with one another. Biological ontologies are supposed to solve the "data silo" problem by creating *controlled vocabularies* for the sharing of biological information.

In the late 1990s, as the number of organisms with completely sequenced genomes grew, several senior scientists in charge of managing the genome databases for these organisms began to realize the need for a shared language. In particular, they needed a way of talking about the functions of genes, the central objects of biological interest. Several efforts had been made to create functional classification systems, but these "were limited because they were not shared between organisms."[15] Suzanna Lewis, who was in charge of FlyBase (the genome database for *Drosophila* fruit flies), reported on some of her emails from 1998:

> Our correspondence that spring contained many messages such as these: "I'm interested in defining a vocabulary that is used between the model organism databases. These databases must work together to produce a controlled vocabulary" (personal communication); and "It would be desirable if the whole genome community was using one role/process scheme. It seems to me that your list and the TIGR [The Institute for Genome Research] are similar enough that generation of a common list is conceivable (personal communication).[16]

In July 1998, at the Intelligent Systems for Molecular Biology (ISMB) conference in Montreal, Michael Ashburner suggested a simple, hierarchical, controlled vocabulary for gene function. His paper, "On the Representation of 'Gene Function' in Databases," was not well received—most participants considered it naïve. Afterward, however, in the hotel bar, Lewis (representing FlyBase) met with Steve Chervitz (representing the yeast database, Saccharomyces Genome Database) and Judith Blake (Mouse Genome Informatics) and agreed to a common scheme for describing the functions of genes.[17] This collaboration became the Gene Ontology (GO) Consortium.

Ashburner was already aware of the work on ontologies in artificial intelligence and medicine, and his proposal included the suggestion that the consortium's gene ontology (GO) be compatible with, or at least translatable to, more generalized ontologies.[18] He argued that the GO had to be more sophisticated than a mere list of terms: "The advantage of a structured graph over a flat keyword list is that you could have a representation of part and whole, it's easy to maintain, it's easy to use, and you have information built into the structure of the graph, whereas with flat keywords, there is only one thing you can do with that: sort it alphabetically."[19] A structured graph looks somewhat like a flow diagram in which terms are linked together by arrows. Figure 4.2 shows a small, simplified section of the GO structured graph. The arrows show how the GO terms are logically linked to one another—in this case, they show that a cell comprises various parts and subparts. "Cell" has two parts, "cytoplasm" and "nucleus." "Nucleus," in turn, has three subparts, "nucleolus," "nucleoplasm," and "nuclear membrane." This structure can be represented in a computer file in XML format with the nodes as GO terms and the connecting arrows as GO relationships. A stripped-down version might look like this:

```
<go:term>
<go:name> cell </go:name>
</go:term>
<go:term>
<go:name> nucleus </go:name>
<go:part_of> cell </go:part_of>
</go:term>
<go:term>
<go:name> nucleolus </go:name>
<go:part_of> nucleus </go:part_of>
</go:term>
<go:term>
<go:name> nucleoplasm </go:name>
<go:part_of> nucleus </go:part_of>
</go:term>
<go:term>
<go:name> nuclear_membrane </go:name>
<go:part_of> nucleus </go:part_of>
</go:term>
```

Using the <go:part_of> tag, a simple computer program (a parser) could read off the hierarchy from this XML. In other words, it could

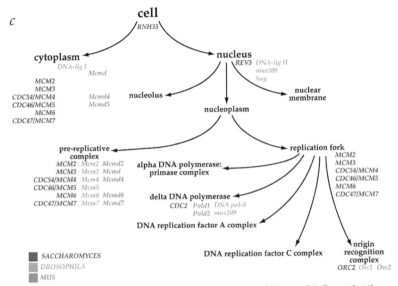

FIGURE 4.2 A representative section of the Gene Ontology (GO) graph. The graph indicates that the cell contains two parts, the cytoplasm and the nucleus; the nucleus, in turn, contains three parts, the nucleolus, the nucleoplasm, and the nuclear membrane; and so on. MCM2, REV3, and so forth are genes associated with these various parts. (Gene Ontology Consortium, "Gene Ontology." Reproduced by permission of Nature Publishing Group.)

understand that the nucleolus is a part of the nucleus and the nucleus is part of the cell, and so on. This structure of the cell and the relationships between its parts are programmed into the GO itself: the GO is a specific structure designed to represent particular features of biological objects.[20]

Since 1998, the GO has evolved into three distinct hierarchies, one describing biological processes (what process a gene product is involved in), one describing molecular function (what molecular mechanisms are involved), and one describing cellular components (where in the cell the gene product acts). The explicit aim is the unification of biology through a shared language: "All biologists now acknowledge that there is likely to be a single limited universe of genes and proteins, many of which are conserved in most or all living cells. This recognition has fueled a grand unification of biology; the information about the shared genes and proteins contributes to our understanding of all the diverse organisms that share them."[21] The GO is designed to be flexible, to dynamically adjust to changing ideas in biology, and to be responsive to the biologists who are using it.[22] Nevertheless, the GO requires a centralized group of "cu-

rators" or "editors" who have a broad knowledge of the overall structure, scope, completeness, and consistency requirements of the ontology
and maintain ultimate control over its terms.[23]

The GO has certainly not solved all communication and consistency
problems in biology. This is, in part, because it was designed only to provide a language for talking about gene function. The success of the GO
has inspired a host of ontologies in other biological domains: cell types,
descriptions of environment, experimental techniques, human diseases,
anatomy, pharmacogenomics, imaging methods, pathways, and human
phenotypes all have their own ontologies. The Open Biomedical Ontologies (OBO) Foundry attempts to bring order to this proliferation
by setting up rules and standards for the creation of ontologies themselves, with the ultimate aim being "a suite of orthogonal interoperable
reference ontologies in the biomedical domain."[24] In particular, OBO
has developed a "relationship types" ontology that specifies the logical connectors ("is_a," "part_of," "has_part," "has_agent," etc.) used by
ontologies.[25] Such ontologies of ontologies, or meta-ontologies, provide
a framework for what their creators hope will be an all-encompassing
language for describing biology and medicine.

Proponents of ontologies believe that they are the only reliable way
of promoting the free exchange of experimental data in science. To make
this point, Barry Smith (one of the founders of the National Center for
Biomedical Ontologies) contrasts ontologies with "folksonomies" of the
kind enabled by the photo-sharing website Flickr.[26] Photos in Flickr can
be "collaboratively categorized" using "freely chosen keywords" based
on the tags that individuals apply to their photos.[27] How is it possible to
choose among the myriad of individual sets of classifications if there are
no constraints? Smith examines four possibilities: deferring to the authority of a "terminology czar," using the first that comes along, using
the best, or using an ontology based on "reality, as revealed incrementally by experimental science."[28] An "instrumental" ontology, based on
a particular model of how biology works, will sooner or later be superseded or outcompeted. A "realist ontology," Smith argues, is the only
way to make ontologies work. Terms in an ontology should correspond
to "what one actually sees in a lab, not what is convenient"[29]:

> Ontology, as conceived from the realist perspective, is not a soft-
> ware implementation or a controlled vocabulary. Rather, it is a
> theory of reality, a "science of what is, of the kinds and struc-
> tures of objects, properties, events, processes and relations in
> every area of reality."[30]

In other words, ontologies are not static lists, but rather dynamic structures that evolve with scientific ideas. Building ontologies, Smith concludes, should be like building scientific theories. The vision for biological ontologies is that they will enforce agreement about what objects exist and the relationships between them—objects in the biological world will have to be fitted into the structure of the ontology.

Some social scientists have taken for granted the important role that ontologies should play in encouraging collaboration and data sharing. Sabina Leonelli argues that "bio-ontology consortia function as a much-needed interface between bottom-up regulations arising from scientific practice, and top-down regulations produced by governmental agencies. They achieve this by focusing on practical problems encountered by researchers who use bioinformatic tools such as databases."[31] She suggests that the centralization of power in consortia like the GO Consortium actually promotes epistemic pluralism in biological research.

My ethnographic experience, however, suggests that ontologies in general, and the GO in particular, are not universally accepted tools among working biologists. This is best appreciated by noting that the GO has several competitors; although the GO is the most widely used gene ontology, biologists often run their data against multiple gene ontologies before being satisfied that the result is plausible.[32] At a lab meeting, one principal investigator warned his students and collaborators: "GO is useless, I always ignore it . . . it's way too general." Expressing her distrust of the GO, another bioinformatician reminded me that ultimately the system was maintained by "a bunch of postdocs" who just sat in a room reading scientific papers and deciding what terms to include. When I raised the subject of ontologies among biologists, they often reminded me of the quip, sometimes attributed to Ashburner, that "biologists would rather share their toothbrush than share a gene name."

Such reactions suggest a profound discomfort among biologists with the centralization of responsibility and the structuring of knowledge that ontologies impose. Biologists often describe their discipline as "messy" compared with sciences like physics or chemistry; what is interesting is found in the exceptions rather than the rules, and careers have been built on the ability to navigate this uniqueness. This view suggests why the kind of standardization that ontologies offer is not always well received: biologists consider the freedom and flexibility of their categories and their language to be an advantage for investigating and describing biological systems. Conforming to standards, even if they are collaboratively developed, and speaking in controlled vocabularies may not be in a biologist's self-interest, at least not in the short term.

The examples of "centralizing" and "democratizing" regimes I have described here entail particular political and technical visions of biology. Each recognizes the need for maintaining a balance between flexibility (in order to promote scientific innovation) and structure (in order to allow data sharing). Where they crucially differ is in their visions of how biological expertise should be distributed in space. Ontologies imagine a grand unification of biology powered by a reduction of language to a universal and machine-readable form—one of the aims of the ontologists is making the GO and other bio-ontologies compatible with OWL (Web Ontology Language). Biological knowledge will be produced largely from the resources of the Semantic Web, guided by a few experts in central locations.[33] The alternative, as exemplified here by DAS, is a wiki-biology—multiple visions and multiple languages will be allowed to flourish and compete for attention and certification. Biological knowledge will be distributed and heavily reliant on local expertise.

The GO attempts to constrain the shape and form of bioinformatic objects—it tries to determine the kinds of things that can exist in digital biology. But it also polices the relationships between them—it has consequences for biological knowledge because it establishes structures and hierarchies through which biological things can relate to one another. As we will see with respect to databases in the next chapter, the structures of information technologies exert powerful forces on the ways in which biologists think about organisms. Simultaneously, the GO has consequences for the disciplinary structure of biology—it establishes hierarchies among different groups of biologists. The GO shows (as we have seen elsewhere) that bioinformatics is a transformation of objects, knowledge, and the organization of biological work.

In the short term, biology will continue to rely on ontologies to promote data sharing. My aim here has been less to criticize such ontologies than to show how specific technical solutions determine specific structures within which biologists must talk and act. Not only the movement of data, but ultimately the authorization of bioinformatic knowledge, depends on the organization and hierarchies within the biological community. But the DAS is an alternative techno-social solution to the problem of data sharing. It suggests that the structures imposed by ontologies are not necessary but contingent—they are built by practicing biologists. Indeed, much of the work of bioinformatics is in generating these ways and means of allowing data to move around frictionlessly.

Another important example of this type of standardization is the work of the Genomic Standards Consortium (GSC). Since 2005, the GSC has attempted to extend the reach of standard vocabularies to cover the provenance of DNA samples. For instance, their Minimum Information about a Genome Sequence (MIGS) creates standards for reporting the geographic location at which a sample was collected, the time, the habitat (including temperature, light, pressure, pH, etc.), health of the organism, sequencing method, assembly method, extraction methods, standard operating procedures, and a range of other factors (all with associated standard vocabularies).[34] To produce the information to meet such a standard, biologists would need to follow specific procedures and methods for gathering and processing DNA samples. That would mean measuring temperature and pH, using a particular assembly method, creating standard operating procedures, and so on. The standardization of language enforces particular ways of working and doing.

In attempting to standardize language, the GO is also flattening biology. It takes the multiplicity and complexity of biological language and renders it into a data structure. The GO shows how bioinformatics depends on the standardization and data-ization of biological objects. But the structure that the GO creates does not affect only computers; it also affects how biologists think about biological objects and what they do with them. A standardization of terms contributes to a standardization of biological practice and biological knowledge. Like other technologies of virtualization, ontologies make biology more compatible with computing by reducing it to standard forms that can be coded, digitized, and shared through electronic networks.

Virtual Spaces

What happens to data once they enter the virtual realm? How do they get manipulated there? This section tracks how data are moved around in the performance of bioinformatic knowledge making. The flattening and virtualization of samples and language into data allow the organization of data in space to become a kind of knowledge production. The value of bioinformatic work consists of this careful spatial ordering according to the structures of databases, data formats, and ontologies. Formatting, managing, and curating data are ways of arranging them into knowledge.

What do the virtual spaces of bioinformatics look like? In the most literal sense, they are text on a computer screen. Most bioinformatics begins with a "Unix shell"—a command prompt at which the bioinformatician can enter commands in order to navigate around. In the first instance, one might only need to navigate around the files stored on one's own computer using such commands as "cd" (change directory) and "ls" (list, which provides a list of all the files and other directories in the current directory). On almost all computers, files are accessed by means of a hierarchical tree of directories, which can be navigated by traveling up and down the branches using the "cd" command. In keeping with the tree metaphor, the topmost directory is usually known as "root" (labeled "/" in Unix). However, almost anything useful in bioinformatics will require accessing files on other computers; these might be the computers of colleagues sitting next to you, they might be networked hard drives sitting in the corner of the lab or the basement of the building, or they might be servers at some distant location. In many cases, the physical location of the machine being accessed makes little or no difference and is unknown to the user. The most common way of connecting to another computer is to use "ssh" or "secure shell"—this is a system of communication that allows you to log into another computer remotely using a username and password. For example, to log into a computer called "tulip" at the EBI, I might type "ssh tulip. ebi.ac.uk." Tulip will prompt me for a username and password. If I am authorized to access tulip, the command prompt will reappear—the screen and the prompt may look exactly the same as if I were using my own computer, and I can now navigate around the tulip machine using exactly the same commands.

This sounds straightforward enough. However, access to such virtual spaces is highly regulated. At the EBI, access to all physical spaces is controlled by RFID cards—when I arrived, I was provided with an ID on the first day. Access to virtual spaces is limited by username and password combinations; by contrast, it took over a week to arrange my access to all the commonly used computers. Moreover, not all computers are directly accessible—sometimes a series of logins is required, ssh-ing from one's own computer to computer A and then from A to B. Some computers can be accessed only by programs (those usually used for intensive calculation), some are dedicated to particular kinds of data (databases, for instance), some are for everyday use, some for use by particular users or groups, some for long-term storage, some for backup, and some for hosting publicly accessible websites. Figure 4.3 gives a sense of the variety of machines involved and the complicated way in which they can be connected to one another. Bioinformaticians use metaphors of space in talking about how they move around these extended networks: tunnels, firewalls, routers, shells, and transfers all suggest a space in which both the user and the data can move around. As I learned to how to log into various machines and find my way around the network, my questions would invariably be answered with

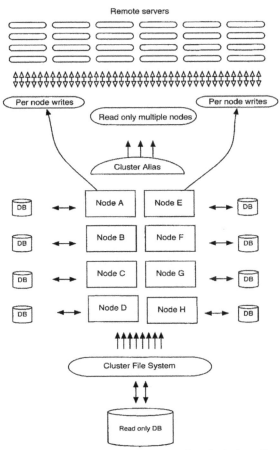

FIGURE 4.3 Computer network conceptual map, European Bioinformatics Institute, Hinxton, UK. Bioinformatics workers need not know the details of this complex physical architecture—the links in virtual space are far more important. (Cuff et al., "Ensembl computing architecture." Reproduced with permission of Cold Spring Harbor Laboratory Press.)

diagrams of boxes (representing computers) connected by lines or arrows (representing potential ssh connections between them). Although bioinformaticians interacted with such systems only textually, such movement relied on a latent mental image of how such machines were virtually linked together.[35]

What are the contents of such virtual spaces? Virtual spaces are inhabited by four types of objects. First, they contain data, either in files or in databases. Second, they contain computer programs (software), which might, at any given time, actually be using or manipulating data. Third, they contain directories, which, along with the databases, are the organizing structure for both the data and the software. Fourth, they may contain bioinformatician-users moving around in the space and manipulating data, software, or the directory structure. The way in which these four elements interact with one another is both intricate and tightly constrained. For instance, beyond the necessity of logging into any machine, certain kinds of data may be manipulated only by certain individuals (often, those who created them). Further, users may not store data just anywhere or execute software on any computer: data must be placed in particular directories or directory structures, and programs must be run on dedicated machines.

The "farm" system in place at the EBI illustrates the care that is taken to control and monitor movement through the virtual space. The "farm" or "compute farm" is a set of computers (549 in November 2008) used by the EBI for running programs requiring intensive calculation or data manipulation. For most users, it is not possible to log into the farm directly. Instead, bioinformaticians log into a computer called "farm-login," which is connected to the farm. In order to run a computationally expensive program, users must submit a "job" using a system called LSF. This program allocates the job to one or more of the machines in the farm. According to the internal wiki:

> All users of the farm initially have equal priority. The queueing system uses a policy called "Fair-share" to ensure equal distribution of work. LSF will dynamically change your priority depending on how much time you have recently consumed on the farm. If you have been running lots of jobs, LSF will lower your priority to allow other users' jobs to be dispatched in preference to yours. Conversely, if you have not used much farm time, LSF will increase your priority with respect to other users.[36]

The farm is designed to "farm out" the computational work in an efficient and equitable manner. The metaphor of working the land is appropriate—if the total network of computers is imagined as a space, it is also imagined as one in which certain resources (disk space, computational power, memory) are valuable commodities that must be tightly regulated. Regulating space allows its productivity to be maximized.

What are bioinformaticians doing in virtual space? At the most literal level, they are usually interacting with screens of text—typing commands, writing programs, and moving or copying files. Almost always, they will be in more than one place at once—there is no limit to the number of computers to which a user can connect via ssh, and bioin-formaticians are routinely logged into many at once, using a window-based operating system to flick back and forth between the various connections. For instance, a bioinformatician's task might involve si-multaneously writing a program on his or her own machine, looking up a database on a public server, and copying data to or from disk space on a third machine. A large part of the virtuosity of such work is in be-ing able to move oneself and one's data rapidly between places. Indeed, bioinformaticians, like software engineers, constantly seek to reduce the number of keystrokes necessary to move around. They can do this by setting up aliases, short commands that act as abbreviations of lon-ger ones. Or they can use their knowledge of programming languages such as Perl and regular expressions to find a shortcut for all but the most intricate of maneuvers. In programs and on the command line, it is common to see bioinformaticians using abstruse strings (for instance: "{^(?:[^f]|f(?!oo))*$}") in order to save themselves extra typing. Hav-ing a working grasp of such intricacies, combined with a knowledge of where important files and programs are located on the network, makes a skillful bioinformatician.

Much of the work of bioinformatics can be understood as the move-ment and transformation of data in virtual space. At EBI, I closely fol-lowed the progress of the "release cycle," a process that occurs every couple of months through which the EBI's main database (known as Ensembl) is revised and updated. A detailed description of the release cycle will illustrate the importance of space management in bioinfor-matic work.

Much of the work of the release coordinator is making sure that the right data end up in the right place in the right form. Ensembl does not produce its own data; instead, its role is to collect data from a wide variety of sources and make them widely available in a common, coher-ent, and consistent format. Ensembl is also an automatic annotation system: it is software that takes raw genomic sequence and identifies the locations of particular structures (e.g., genes) and their functions. Mak-ing a release involves collecting data from many individuals and places and running them through a software pipeline. For such a large set of databases, it is not possible to simply update them one by one. Ensembl requires a sophisticated "staging" system whereby the new release is prepared, processed, tested, and checked for consistency before it is re-leased "live" onto the World Wide Web. Thus the release cycle becomes a carefully choreographed set of movements through a virtual space in which data are migrated from the "live mirror" to the "staging server" and back again. I have excerpted the following instructions from the technical document used by the release coordinator:

> Firstly, check there is enough space for the new databases, e.g. in /mysql/ data_3306/databases/df -h /mysql
> # Release 50: 1000G 342G 658G 35% /mysql
>
> You should be able to see how much space is required here (may be out of date):
> http://www.ensembl.org/info/webcode/requirements.html http://www .ensembl.org/info/webcode/installation/index.html
> Release 50: Total required = 632.8 GB
>
> Next you will need to modify the following script:
> cd /ensembl-personal/release_coordination/scripts
> . . .
> Now, run the script. For release 50:
> cd /ensembl-personal/release_coordination/scripts
> perl get_databases_from_ens-livemirror_to_ens_staging.pl 50
>
> The script generates an appropriate input file, ens-livemirror_to_ens_staging_databases to use with ensembl/misc-scripts/CopyDBoverServer.pl
>
> To actually copy the databases across from ens-livemirror to ens-staging you must be on ens-staging and logged in as mysqlens. Ask a previous release coordinator for the password if you don't already know it.
> ssh ens-staging
> su—mysqlens
>
> Now run the copy script:
> cd ensembl/misc-scripts/
> perl CopyDBoverServer.pl -pass ensembl /path/to/file/ens livemirror_to_ens_staging_databases
>
> Save and check the output from this script.[37]

These instructions describe in detail the procedures used to copy databases from the "live mirror" to the "staging server." The crucial knowledge in this document is about *where* to find the various databases and the programs to move them. The "scripts" are short Perl programs (for example, get_databases_from_ens-livemirror_to_ens-staging.pl— the purpose of the program is made obvious by its name) that are used as shortcuts for copying large numbers of databases at once. In order to achieve the last step, the release coordinator must be "on" (that is, logged into) the staging server. The document, over forty pages long in all, provides step-by-step instructions on how to move oneself and the data around in virtual space order to perform the release cycle.

This description of the work at EBI shows the complexity of collecting, storing, and organizing biological data. Without such work, genomic biology would not be possible. Ensembl is an essential tool for managing big data, for making long strings of As, Gs, Ts, and Cs into "genes," "regulatory elements," and other biological objects. The Ensembl web-based interface attracts hundreds of thousands of users per month, and it has been cited in hundreds of research articles.[38] User-biologists can download short sequences or whole genomes, access specific data based on Ensembl's annotations (for instance, just find particular genes), compare data between genomes, analyze intraspecies variation, or examine regulatory sequences. What biologists can do with Ensembl is exactly the kind of thing biologists need to do all the time: it is the very work of biology itself. For example, the following excerpt comes from a website giving advice about how to use Ensembl:

> Let's say you have a set of genes in one species and you want to know the orthologs in another species and gene expression probes in that species you can use to assay those orthologs. For example, [given] 25 gene expression probes that are dysregulated in humans when exposed to benzene. What if you only had the U133A/B Affymetrix probe IDs and wanted to know the gene names? What if you also wanted all the Ensembl gene IDs, names, and descriptions of the mouse orthologs for these human genes? Further, what are the mouse Affymetrix 430Av2 probe IDs that you can use to assay these genes' expression in mouse? All this can be accomplished for a list of genes in about 60 seconds using [Ensembl].[39]

Ensembl's tools allow biologists to deal rapidly with large amounts of data in different formats, comparing them across different organisms. Ensembl and other tools like it are the most valuable resources available for studying genes and genomes, allowing biologists to manipulate and analyze the vast amount of available genomic data.

Bioinformatics is the task of ordering biological data in order to make them usable for biological research. Bioinformaticians must strive to maintain close control over their spaces, restricting access and protecting hierarchies and structures, because it is the integrity of this space that determines the value of their work—in other words, it is through the organization of data in virtual space that bioinformation can become bioknowledge. Ensembl's work should be understood as integral to the knowledge-making practice of biology. Data management or data curation consists of this work of organizing data into value. A disordered or haphazardly organized space would be of no value because it would contain no knowledge, reveal nothing about biology. Thus the ways data are moved around in space and the ways they are arrayed in space determine their epistemological status—data that are appropriately ordered in virtual space can attain the status of biological knowledge.[40]

A great deal of effort and computational time is invested in keeping the data where they should be because the Ensembl brand depends on it: if information in the publicly accessible databases is to maintain its reputation for being reliable and usable knowledge, it must remain highly ordered at all times. The Ensembl team itself conducts no experiments, produces no raw data directly from wet samples. Yet its work is highly valued by biologists because of its contribution to organizing data into knowledge. The particular ways in which Ensembl structures and organizes information are considered to be its trademark and its greatest asset.

The careful management of and navigation through virtual space constitutes the work of bioinformatic knowledge making. The flattening of both sequences and language described earlier in this chapter makes it possible to do this kind of biology in virtual space. It does not, however, make biology completely fluid. Rather, the constitution of biological objects becomes dependent on all sorts of structures built into hardware and software. By understanding networks and hard drives as a space that can be traversed, surveyed, farmed, and mapped, we can begin to understand how biology is constrained by the virtual spaces it inhabits.

Conclusions

This chapter has considered the virtual spaces in which biologists move around and through which they make knowledge. In contemporary biology, different kinds of spaces and different kinds of movements through them produce different kinds of knowledge. Information technologies require the flattening of biological objects and language into data. Pipelines and ontologies are both ways of achieving this flattening.

These structures open up new modes of knowledge creation through the movement of data in virtual space. But the kind of knowledge that emerges is dependent on and shaped by the hardware and software it subsists on: data structures, directed graphs, databases, servers, and compute farms.

These processes of flattening are also processes of making data fit for travel. The sequencing "pipeline," for instance, produces a particular kind of bioinformatic knowledge that is stripped of its messiness, contingency, and situatedness; this process provides sequences with their peculiar ability to travel and recombine as fluid, universal objects. Much has been made of the fact that bioinformatic biology is globalizing. Many accounts suggest that biological objects are becoming place-free, homogeneous, and fluid. But as far as biological work goes, this uniformity and liquidity is not automatic—as this chapter has shown, it is an achievement, a consequence of biological work. The pipelines and ontologies and release cycles are ways of making heterogeneous and lumpy data into smooth, universal data that can travel with ease.

In other words, this process is making the local appear universal. Understanding how biology has become and is becoming global or universal must begin with understanding how certified knowledge is produced in the labs and networks such that it can then travel through global spaces. Biotechnology, personalized genomics, and so on seem to be playing out on a worldwide scale; however, what is significant in these fields is how knowledge is made such that it *can* be globalized—how is it that a sequence produced at 320 Charles becomes *the* sequence of an elephant's genome? In this chapter we have seen that it is movement through particular sorts of spaces, and in particular, movement through the "sequencing pipeline," that constitutes this sort of knowledge production: the transformation from a particular material sample to a universal (and universally certified) digital representation takes places through a detailed and carefully policed series of steps acted out in space. The feature of globalization relevant to bioinformatics is not deterritorialization, but rather interconnectedness: the specificity of particular situations and conformations in particular laboratories and networks has a greater ability to make their influence felt everywhere.[41] The Broad's elephant becomes everyone's elephant. In assessing the status and significance of globalized biological knowledge, we need to continually recall its dependence on the situations from which it springs. Doing so will remind us that bioinformatic knowledge is never automatically fluid or universal, but always dependent on carefully constructed networks, structures, hierarchies, and sets of movements that keep it mobile.[42]

The production of biological data entails a particular kind of working and a particular kind of knowing. Central to this point is the idea that bioinformatics requires standardization—a kind of flattening of the biological object—in order to function. Moving objects around in virtual space means making them computable, networkable, and so on. Nucleotide sequences have become such standardizable, computable, networkable objects. Making bioinformatics, then, has had much to do with constructing sequences in this way—as just such standard objects. The reducibility of sequence to data, to objects that can flow through the computer, has played a major role in establishing its importance and ubiquity in contemporary biological work. Sequence permits precisely the kind of abstraction or stripping down that is required for samples to be transformed into data. Bioinformatics has emerged "out of sequence" because it is sequence that has made it possible to move biology around virtual space.

5 Ordering Objects

Of all the structures that computers impose, databases are the most important. If we wish to understand classification and its consequences in the late twentieth and early twenty-first century, we need to understand databases and their role in the making of scientific knowledge. In biology, the influence of databases on practice and knowledge is profound: they play a role in data analysis, transmission, and communication, as well as in the verification and authentication of knowledge. In their day-to-day work, many biologists use databases for checking the results of their experiments, for storing and managing their data, or for performing simulations and experiments. How are such databases built, and by whom? How do they work? What kinds of structures do they contain? How do these structures influence the knowledge that can be made with them? This chapter explores the role of databases in scientific knowledge making using one prominent example: GenBank.

Biological databases, organized with computers, cannot be thought of as just *collections*.[1] Instead, biological databases are *orderings* of biological materials. They provide ways of dividing up the biological world; they are tools that biologists use and interact with. Computer databases store information within carefully crafted digital structures. Such tabulations can have profound social, cultural, political, and economic consequences.[2] But as well as ordering society, databases construct orderings of scientific knowledge: they are powerful classification schemes that make some information accessible and some relationships obvious, while making other orderings and relationships less natural and familiar.[3] Organizing and linking sequence elements in databases can be understood as a way of representing the connections between those elements in real organisms. Like a billiard-ball model of a gas in physics, databases do not aim to be a straightforward representation of a biological system; rather, they aim to capture only some of its important features. The database becomes a digital idealization of a living system, emphasizing particular relationships between particular objects.

As I worked with biological databases in my fieldwork, I started to ask why the information in them was arranged the way it was. Indeed, how did databases become the preeminent way of storing biological data? Answering these questions required an interrogation of the history of databases. By examining a database diachronically, we can discover how changes in structure correspond to changes in the kind of work being performed (and in the knowledge being produced) through databases.

The different database structures that GenBank has used represent different ways of understanding and ordering biological knowledge. Early "flat-file" databases, such as those constructed by Margaret Dayhoff and the first iterations of GenBank, instantiated a protein-centered view of life in which single sequence elements were placed at the center of biological understanding. The "relational" databases that gradually replaced the flat files in the 1980s and 1990s emphasized the interconnections between sequence elements—biological function was produced by interaction between different elements, and the connections were reflected in the database. Finally, the "federated" databases of the postgenomic era, while still placing sequences at the center, allowed much wider integration of other (extra-sequence) data types. This gave structural expression to the notion that biological function could be best understood by modeling the relationships between genes, proteins, transcription factors, RNA, small molecules, and so on. By following data into databases, we see how the rigid structures of information technologies impose constraints on how data can move and be shaped into knowledge.

The activities of data storing and knowledge making are not separate and are not separable. Biological databases are not like archives and museums—they are oriented toward the future more than the past.

Treating them as part of "natural history" can cause us to overlook their importance in structuring the way biologists make knowledge about life. The computer is more than a mere organizing tool or memory device—it provides ways of representing, modeling, and testing biological systems. The sort of biological databases that arose in association with computers in the 1960s marked a new kind of object and a new, "theoretical" way of doing biology.[4]

A Brief History of Databases

Databases have a history independent of their use in biology. Like the computer, they were built for specific purposes for the military and for business data management. How do they work? What were they designed to do? The first databases—or data banks, as they were often called—were built, like so many other tools of the information age, for military purposes. Thomas Haigh argues that the Semi-Automatic Ground Environment (SAGE) was the first "data base." SAGE needed to keep track, in real time, of the status of bombers, fighters, and bases in order to serve as an automated early warning and coordinated response system in the event of aerial attack on the United States.[5] In the early 1960s, SAGE's creators at Systems Development Corporation were actively promoting "computer-centered data base systems" to business. The corporate world soon took up the idea of a management information system (MIS), which many hoped would provide an executive with instant access to all the pertinent information about his organization. Early MISs were essentially file management systems—pieces of software that contained generalized subroutines to open, close, and retrieve data from files. This technology was limited, however, by the fact that data records were stored one after another along a tape, and that to find a particular record, it was often necessary to scroll through large portions of tape. The introduction of disks in the early 1960s meant that data could be accessed at "random," and new possibilities arose for organizing data access and storage.

Beginning in 1963, Charles W. Bachman of IBM developed the Integrated Data Store (IDS), which became one of the most effective and influential file management systems. The IDS, designed for use with disks rather than tapes, allowed linkages between records in what came to be known as the "network data model." To find particular records in such a system, the user had to navigate from record to record using the various links.[6] A version of the IDS was used to run computers for the Apollo program in the late 1960s.

In 1970, Edgar F. Codd, working for IBM in San Jose, California, wrote a paper describing a new system for organizing data records. "A Relational Model of Data for Large Shared Data Banks" set out a scheme by which the representation of data could be completely divorced from their physical organization on a disk or tape: "Activities of users at terminals and most applications programs should remain unaffected when the internal representation of the data is changed. . . . Changes in data representation will often be needed as a result of changes in query, update, and report traffic and natural growth in the types of stored information."[7] Codd's idea was to organize the data into a set of tables that were related to one another by "keys" that linked data across tables.[8] For instance, a library database might contain information about all the books in its collections. Such information could be spread over multiple tables, as in this example (a library system with just four books):

BOOK_ID	AUTHOR_ID	BOOKNAME	CALL_NO	LIBRARY_NO
1	1	Making Sense of Life	XYZ	1
2	2	Simians, Cyborgs, Women	ZYX	2
3	2	Primate Visions	YXY	1
4	3	Nature and Empire	YYZ	2

AUTHOR_ID	FIRST_NAME	LAST_NAME
1	Evelyn	Keller
2	Donna	Haraway
3	Londa	Schiebinger

LIBRARY_NO	LIBRARY_NAME	LIBRARY_ADDRESS
1	Library of Congress	Washington
2	New York Public Library	New York

Columns with identical names in different tables ("AUTHOR_ID" and "LIBRARY_NO") are linked together in the database. To find the author of *Primate Visions*, for example, the database must first look up that title in the first table, retrieve the AUTHOR_ID (the "key" for the table of authors), and then look up the corresponding author in the second table. The query itself creates a link or "join" between two tables in the database. Codd's paper also suggested a "universal data sublanguage" that could be used to query and update the database. Such a language would refer only to the names of the tables and the names of the columns within them, meaning that a command or query would still work even if data were reorganized; as can be seen in the library example, the rows and the columns could be rearranged without affecting the outcome of a query.

The relational model had two advantages over its "network" rivals. First, it did not require relationships between data to be specified during the design of the database; second, the abstraction of the structure from the physical storage of the data greatly simplified the language that could be used to manipulate the database. "Because the relational model shifted the responsibility of specifying relationships between tables from the person designing them to the person querying them," Haigh argues, "it permitted tables to be joined in different ways for different purposes."[9] Relational databases present an open-ended, flexible, and adaptable means to store large amounts of data. Despite IBM's initial support for the "network" model, the development of Codd's ideas through the 1970s led to the development of SQL (Structured Query Language) and the commercialization of the relational model through firms such as Oracle and Sybase.

Even from this brief history, it is clear that different types of database structures are appropriate for different types of data and for different types of uses. Moreover, this history suggests that databases act as more or less rigid structures for containing information—that the proximity and accessibility of particular kinds of data are determined by the form of the database itself.

Dayhoff and a New Kind of Biology

The first biological databases—that is, the first groupings of biological information ordered on a computer—were produced by Margaret Oakley Dayhoff. Dayhoff, born in 1925, was trained in quantum chemistry under George E. Kimball at Columbia University, receiving her PhD in 1948. Her thesis work involved calculating the molecular resonance energies of several polycyclic organic molecules—a computationally intensive problem that involved finding the principal eigenvalues of large matrices.[10] In approaching this problem, Dayhoff devised a way to use punched-card business machines for the calculations. After her graduate studies, Dayhoff pursued her research at the Rockefeller Institute (1948–1951) and at the University of Maryland (1951–1959). In 1960, she joined Robert Ledley at the National Biomedical Research Foundation (NBRF, based at Georgetown University Medical Center, where she also became a professor of physiology and biophysics), and it was here that she turned her attention to problems of proteins and evolution. Ledley himself was a pioneer in bringing computers to bear on biomedical problems, trained as a dentist, but also as a physicist and a mathematician, during the early 1950s, Ledley worked with the Standards Eastern Automatic Computer at the National Bureau of Standards (NBS) in Maryland. The knowledge of digital computing architecture that Ledley attained at the NBS led him first to problems in operations research (OR) and then to the application of computers to biomedicine. In particular, Ledley was interested in using computers to create a mathematized biology that would allow, for example, computerized medical diagnosis.[11]

Ledley had a very specific vision of how computers would be useful to biology. In his OR work, Ledley had emphasized the translation of messy situational data into logical problems that computers could understand and solve. In his work with George Gamow on the genetic code, Ledley devised a way for biologists to translate their protein-coding schemes into matrices and symbolic logic that could be easily dealt with on a computer.[12] Likewise, in biology and medicine, computers would be tools that could be used for statistics, accounting, and data management.[13] In his lengthy survey of the field (published in 1964, although much of it was written some years earlier), Ledley outlined his justification for the computer management of biomedical information:

> The feasibility of such a system from a computer-technology point of view is unquestioned; there are already computers that carry out such closely related processes as making nation-wide airline and hotel reservations, recording, updating, and tallying bank accounts and other financial records, controlling large-scale defense installations, and so forth.[14]

At the NBRF, Dayhoff and Ledley began to apply computers to the solution of problems involving large quantities of experimental data. In 1962, the pair developed a computer program to aid in the experimental determination of protein sequences. Previously, the only way to experimentally determine a protein's complete sequence was to find the sequences of short fragments of the chain and then "try to reconstruct the entire protein chain by a logical and combinatorial examination of overlapping fragments."[15] For larger protein chains, it quickly became an almost impossible task to assemble fragments by hand. Dayhoff and Ledley's program not only rapidly checked possible arrangements, but also suggested the best approach for further experiments where results were inconclusive. This program was followed by another, more sophisticated program that allowed for experimental errors and assessed the reliability of the finished sequence.[16] Dayhoff's choice of journal for publishing this work—the *Journal of Theoretical Biology*—suggests that she saw it as making a contribution to the organization and systematization of biological knowledge.

At about this time, in the early 1960s, Dayhoff began to collect complete protein sequences. Her reasons were twofold. First, protein sequences were important in their own right, since they contained the key information about how biology worked. Second, and perhaps more importantly for Dayhoff, proteins contained information about evolutionary history. At the same time that Dayhoff was beginning her collection efforts, Linus Pauling and Emile Zuckerkandl were developing a new method of studying evolution, and the relationships between organisms, using protein sequences as "documents of evolutionary history."[17] Dayhoff and others saw that such work would require both collections of proteins and computer programs to perform the computationally intensive tasks of sequence comparison and phylogenetic tree construction. Dayhoff and her colleagues at the NBRF scoured the published literature for experimentally determined protein sequences and entered those sequences on punched cards for computer processing. Although the collection itself was a nontrivial task, it was never Dayhoff's ultimate aim to be a botanist of sequences: "There is a tremendous amount of information regarding evolutionary history and biochemical function implicit in each sequence," she wrote to a colleague, "and the number of known sequences is growing explosively. We feel it is important to collect this significant information, correlate it into a unified whole and interpret it."[18] Collection was a means to an end.

The first edition of the *Atlas of Protein Sequence and Structure*, published in 1965, listed some seventy sequences. Subsequent editions contained not only the protein sequences themselves but also extensive analyses performed by computer. These analyses included studies of the evolution of specific protein families, the development of a model of evolutionary change in proteins, an analysis of patterns in amino acid alleles, simulation of protein evolution, and studies of abnormal human hemoglobins, ribosomal RNA, enzyme activity sites, and transfer RNA.[19] The *Atlas* also provided phylogenetic trees and protein secondary (three-dimensional) structures. In the preface to the third edition of the *Atlas*, Dayhoff and her collaborator Richard Eck outlined their approach to the sequence collection problem:

The mechanical aspects of the data presentation have been automated. The title, references, comments, and protein sequences
in one-letter notation are kept on punched cards. The alignments, the three-letter notation sequences, the amino-acid com-
positions, the page layouts and numbering, and the author and
subject index entries from the data section are produced automatically by computer.[20]

Although sequences had to be collected and entered from the published
literature by hand, the aim was a computer-ready set of sequence information that could be rapidly subjected to analysis.

Two of Dayhoff's analytical concepts are particularly significant.
First, Dayhoff realized that a careful study of the evolution of proteins
would require a model of how proteins could mutate, and in particular,
which amino acids could be swapped with one another in a sequence
(called a "point mutation"). A naïve approach would treat all such
swaps as equally likely—asparagine could just as easily be swapped for
lysine as for valine, despite the chemical differences between the two
amino acids. If biologists wanted a better account of the evolutionary
distance between sequences, however, a more sophisticated approach
was required. To provide this approach, Dayhoff invented the notion of
a PAM (point accepted mutation) matrix. The idea was to use the protein sequence data she had collected to create a table (or matrix) showing the number of times each amino acid was observed to mutate to
each other amino acid. Dayhoff then computed a "relative mutability"
for each amino acid by dividing the total number of observed changes in
an amino acid by the number of total occurrences of that amino acid in
all the proteins examined. By using the relative mutability to normalize
the mutation data, Dayhoff arrived at a matrix that "gives the probability that the amino acid in column j will be replaced by the amino acid in
row i after a given evolutionary interval."[21] The non-diagonal elements
of the PAM have the values

$$M_{ij} = \frac{\lambda m_j A_{ij}}{\sum_i A_{ij}},$$

where A is the matrix containing the point mutation values, m is the
relative mutability of each amino acid, and λ is a proportionality constant. The elegance of Dayhoff's scheme is that it is possible to simulate different periods of evolutionary time by multiplying the matrix by
itself—a single PAM matrix corresponds to the amount of time in which

each amino acid has a 1% chance of mutation. For instance, multiplying PAM by itself 20 times—often the result is called PAM20—yields a matrix in which each amino acid has a 20% chance of mutating. As Dayhoff was aware, using matrices such as PAM250 can be extremely helpful in detecting distant evolutionary relationships between proteins.

Dayhoff's larger aim was to use such models of mutation to explore the evolutionary relationships among all proteins. From the fifth edition onward (1973), the *Atlas* was organized using the concept of protein "superfamilies," Dayhoff's second major analytical contribution. Families of proteins were already well recognized and easily determined through simple measurements of sequence similarity. The sensitivity of Dayhoff's methods of comparison (using the PAMs), however, allowed her to sort proteins into larger groups, organized according to common lines of descent.[22] Such classifications were not merely an organizational convenience—they provided theoretical insight into the process of evolution. The ultimate aim of the NBRF's sequence collection work was this kind of conclusion:

> In examining superfamilies, one is struck by the highly conservative nature of the evolutionary process at the molecular level. Protein structures persist through species divergences and through gene duplications within organisms. There is a gradual accumulation of change, including deletions and insertions as well as point mutations, until the similarity of two protein sequences may no longer be detectable, even though they may be connected by a continuum of small changes.[23]

The superfamily concept was both a tool of classification and a biological theory. It was a way of conceptualizing the relationships among the entities that made up living things and of making sense of their history. In an article published in *Scientific American* in 1969, Dayhoff outlined some of the conclusions of her work on the classification and history of life: "The body of data available in protein sequences," she argued, "is something fundamentally new in biology and biochemistry, unprecedented in quantity and in concentrated information content and in conceptual simplicity . . . because of our interest in the theoretical aspects of protein structure our group at the National Biomedical Research Foundation has long maintained a collection of known sequences. . . . In addition to the sequences, we include in the *Atlas* theoretical inferences and the results of computer-aided analyses that illuminate such inferences."[24]

Understanding Dayhoff's databasing and collection efforts requires understanding of the computational-theoretical practices in which they were embedded. Although Dayhoff's database was not distributed electronically (it was available on magnetic tape from 1972, but only a handful of tapes were sold[25]), it was stored in computer-readable form, and all the data processing was performed digitally. The *Atlas* was something fundamentally new because it was not just a collection, but provided a system and a means for ordering, classifying, and investigating the living world without doing bench-top experiments. Producing PAMs, defining superfamilies, and generating phylogenetic trees from the sequences were integral parts of the process of producing the *Atlas*. These activities, which were woven into the production and structure of the *Atlas* itself, made it more than a means of collecting and redistributing data; rather, it was a way of organizing, systematizing, and creating biological knowledge.

Bruno J. Strasser argues that Dayhoff's collection efforts (much like botanical gardens of the eighteenth and nineteenth centuries) relied on creating a "network of exchange" or a "Maussian system of gift and counter-gift," but that this system conflicted with "ideas about credit, authorship, and the property of knowledge in the experimental sciences."[26] In particular, Dayhoff's collection and use of other researchers' experimental work (some of it unpublished) conflicted with the dominant norms in biochemistry and molecular biology, in which one's own work was one's own property (particularly if it was unpublished). This conflict manifested itself in several ways. First, it meant that researchers were, by and large, uncooperative—experimenters were reluctant to share their unpublished sequences with the NBRF. Second, Dayhoff had trouble receiving scientific credit for her work. John T. Edsall commented on Dayhoff's prospects for election to the American Society of Biological Chemists:

> Personally I believe you are the kind of person who should become a member of the American Society of Biological Chemists . . . but knowing the general policies that guide the work of the Membership Committee I must add that I can not feel at all sure about your prospects for election. Election is almost invariably based on the research contributions of the candidate in the field of biochemistry, and the nomination papers must include . . . recent work published by the candidate, to demonstrate that he or she has done research which is clearly his own.

> The compilation of the *Atlas of Protein Sequence and Structure*
> scarcely fits into this pattern.[27]

Dayhoff's *Atlas* was considered by some to be nothing more than a mere aggregation of others' work.

No doubt some of Dayhoff's problems stemmed from researchers' reluctance to share unpublished data. At a deeper level, though, this reluctance stemmed from a misunderstanding of Dayhoff's project. As Edsall's attitude suggests, Dayhoff's work was understood as the un-original work of collection and compilation, rather than as an attempt to systematize biological knowledge. Indeed, Dayhoff complained about the "great hostility of journal reviewers" when she tried to present her work as a theoretical contribution to biology.[28] No doubt this had to do with the generally marginal status of theory within biology, and with the prevalent notion that any such theory should look like a mathematical theory in physics, rather than a system of categorization or a database.

Ultimately, after struggling to maintain funding for her *Atlas*, in 1981, Dayhoff and the NBRF failed to win the contract from the NIH to build and maintain a national sequence database (as described in chapter 1, the contract was awarded to Walter Goad at Los Alamos). This failure was a harsh blow for Dayhoff, who had struggled for over a decade to gain support and recognition for her work. The lack of adequate funding had forced the NBRF to charge research biologists a fee for the *Atlas*. This, in turn, embittered the biological community, who saw the NBRF as taking their own work (for free) and selling it for a profit.

The NIH's decision was based on the conclusion that Dayhoff did not have the technical expertise to build and run a modern database.[29] It was Dayhoff, however, who had pioneered the idea of organizing biological data into computerized databases. Although GenBank, as we shall see in the next section, placed a far greater emphasis on using electronic means to collect and communicate data, the notion of using a structured digital space to order biological knowledge and create models of the biological world was Dayhoff's.

Dayhoff created a model for studying evolution. The use of sequence data in conjunction with the PAM matrices and mathematics developed by Dayhoff and her collaborators made it possible to apply evolutionary theory to make specific predictions about the relatedness of species and hence about the history of life. In other words, it was a way of making biological knowledge—without the laboratory or the field—through the structuring and ordering of data.

Dayhoff's principal innovation was not the collection of the sequences, but the use of this collection to investigate biology without doing lab experiments. Because the *Atlas* was largely distributed on paper, this type of investigation was at first mostly limited to the NBRF. As GenBank developed mechanisms for electronic distribution (via magnetic tape and over telephone-based networks), such practices spread.

GenBank

Like Dayhoff's work, the early history of GenBank must be embedded within a culture of practice—databases were developed not just as collections or repositories of data, but as tools for performing specific kinds of biological work. In other words, they were active sites for the development of biological knowledge. An account of the events that led to the creation of GenBank has been given by Temple Smith, who was closely involved with the events he describes.[30] Smith ascribes the advent of sequence databases to the coincidentally simultaneous invention of techniques for sequencing DNA and of mini- and bench-top computers. Although he describes some of the problems encountered by the early databases, he emphasizes that the founders "foresaw both the future needs and the potential of databases."[31]

The advocates of GenBank certainly saw the value of creating a repository for nucleotide sequences in order to manage the output of large-scale sequencing efforts, but they had to do much work to convince potential funders and other biologists of its value. Those actively managing the databases had to make the case that they were far more than collections; they argued that databases should be dynamic structures and tools through which a new kind of biology could be practiced. To most biologists, a database meant little more than an archive, not an important tool for basic research. The caution with which the NIH approached databases led to the construction of a "flat-file" structure for early versions of GenBank. Even this flat-file database, however, had important consequences for how biologists were able to construe and construct the relationships between biological entities.

In addition to Dayhoff's efforts at the NBRF, several other biological database efforts had been inaugurated by the late 1970s. In 1973, protein X-ray crystallographic data collected by Helen Berman, Olga Kennard, Walter Hamilton, and Edgar Meyer had been made available through Brookhaven National Laboratory under the direction of Thomas Koetzle.[32] The following year, Elvin Kabat, an immunologist at Columbia University, made available a collection of "proteins of immu-

nological interest" (largely immunoglobulins) via the PROPHET computer system.[33] By the end of the 1970s, there was sufficient interest in
biological databases to attract about thirty-five scientists to a meeting
on the subject organized by Norton Zinder, Robert Pollack, and Carl W.
Anderson at Rockefeller University in March 1979. A summary of this
meeting circulated the following year within the NIH, listing the reasons why a nucleic acid sequence database was needed:

> 1) the rapidly increasing rate at which nucleic acid sequence
> information is becoming available (approaching 10^6 nucleotides
> per year); 2) the wide range of biological questions that can be
> asked using a sequence data base; 3) the fact that only a com-
> puter can efficiently compare and transform the data base to ask
> questions of interest; 4) the desirability of avoiding a duplication
> of effort in both adding to the data base and analyzing it; 5) the
> desirability of correlating a nucleic acid sequence data base with
> other features of biological importance including mutations,
> natural species variation, control signals, protein sequence and
> structure, nucleic acid secondary and tertiary structure.[34]

For the workshop participants, the main point of the database was to
"ask questions of interest" and to "correlate" the sequence data with
other sorts of biological information. It was not supposed to be an archive or a stand-alone repository. But the applicability of computers,
and particularly computer databases, for asking and answering biological questions was not universally acknowledged. The NIH had not up
to this time funded biological databases, and it had to be convinced
that the effort was worthwhile. The fact that a report of the Rockefeller
meeting took over eighteen months to reach the NIH is perhaps indica-
tive of the priority that it was accorded.

Moves toward a database continued to proceed slowly. Dayhoff,
Goad, Frederick Blattner (from the University of Wisconsin), Laurence
Kedes (from Stanford University Medical Center), Richard J. Roberts
(from Cold Spring Harbor Laboratories), and a few others were push-
ing for the NIH to fund a database effort. In July 1980, Elke Jordan
and Marvin Cassman from the National Institute of General Medical
Sciences (NIGMS) convened a further workshop to discuss prospects
for a database. In contrast to the report of the Rockefeller meeting, the
official report stated only that "an organized effort must be initiated to
store, catalog, and disperse" nucleotide sequence information.[35] Around
the middle of 1980, there was considerable uncertainly as to whether

any federally funded database effort would proceed. Jordan and Cassman received numerous letters supporting the proposed database from
molecular biologists around the country. The correspondence argued
for the database on the grounds that it would act as a powerful organizing resource for biology as well as a repository:

> There appears to be some question as to the utility of a na-
> tional DNA sequence analysis and databank facility. We wish
> to express our strong support in this matter. . . . In our labora-
> tory, we have used Seq [a sequence analysis program available
> at Stanford], for example, to locate transcripts from an *in vitro*
> transcription system when we could not find them ourselves. . . .
> Such a system for DNA sequence analysis would open a new
> way of thinking about sequence analysis for researchers who do
> not now have access to a computing center or staff available to
> maintain a local facility.[36]

The database would not be just a library or an information-sharing
scheme, but provide a "new way of thinking" about sequences for molecular geneticists.

By mid-1980, in order to encourage the NIH to act, both Dayhoff
and Goad had begun pilot nucleotide sequence banks (no doubt they
both also hoped to improve their own chances of winning any NIH
contract that might be tendered). As described in chapter 1, Goad was
a theoretical physicist by training, and after working on nuclear weapons, he became interested in molecular biology in the mid-1960s. At
Los Alamos, he assembled a small group of mathematicians, physicists,
and biologists to work on problems of protein and nucleotide sequence
analysis. Already by December 1979, Goad and his team had written
a proposal for a "national center for collection and computer storage
and analysis of nucleic acid sequences" based on their pilot project. The
aims of such a facility were clearly set out:

> The discovery of patterns inherent in base sequences can be
> aided by computer manipulation to an even greater extent than
> for either numerical relationships (where there is a natural or-
> dering) or natural language text (where we are habituated to
> certain patterns). . . . The library would be invaluable for relat-
> ing sequences across many biological systems, testing hypothe-
> ses, and designing experiments for elucidating both general and
> particular biological questions. . . . The development of methods

capable of answering the most penetrating questions will result from dedicated, ongoing research combining mathematics, computer science and molecular biology at a high level of expertise and sophistication.[37]

By this time the pilot project contained about 100,000 bases. More importantly, though, its sequences were not only embedded within a sophisticated set of programs for performing analysis, but also "arranged in a number of tables for access and manipulation."[38] The team at Los Alamos had adapted a system called FRAMIS, developed by Stephen E. Jones at Lawrence Livermore National Laboratories, that allowed sequence data and the associated biological information to be linked together by sophisticated logical and set theoretic operations.[39] Although this system was difficult to implement (compared with just listing the sequences one after another in a file), the advantage of storing sequence data and other biological information in such a way was that it allowed relationships to be rearranged or information added at a later point without having to alter each individual database entry.

During the late summer and fall of 1980, several unsolicited proposals were made to the NIH. On August 13, Dayhoff requested funds to expand her pilot project; on August 28, Douglas Brutlag, Peter Friedland, and Laurence Kedes submitted a proposal that would turn their MOLGEN project into a national computer center for sequence analysis; on September 3, Los Alamos submitted a revised proposal based on its pilot DNA data bank; and on September 8, Michael Waterman and Temple Smith submitted a supplementary proposal for sequence analysis.[40] The NIH, however, continued to hesitate. Jordan convened a follow-up to the July meeting on October 26. Notes from this meeting made by Frederick Blattner indicate that the decision had been made to segregate the databasing efforts into two separate projects: the first would focus on collection and distribution of the sequence data, and the second on software and more sophisticated analysis tools with which to manage and use these data. Blattner's early sketch of John Abelson's proposed "planning structure" divided the database project between "data collection groups" and "programming": the data collection was to be "annotated, but not sophisticated."[41] The agenda for the third meeting, held in early December, already included a detailed breakdown of the tasks to be performed under the two separate contracts. The scope of work for the first project was to "acquire, check and organize in machine-readable form the published data concerning base sequences in polynucleotides," while the efforts to develop a "database management system . . . for the sequence data that allows sophisticated search capabilities comparable to relational data base" were reserved for the second.[42] Although the NIH intended the two contracts to be contemporaneous and closely connected, by the time the request for proposals was finally made (near the end of 1981), only the first was to be funded.

Dayhoff, Goad, and a small group of other computer-savvy biologists realized that a nucleotide sequence database had to be a sophisticated theoretical apparatus for approaching biological problems. The majority of their colleagues, however, while realizing the importance of a repository, believed that making a database was essentially the trivial process of reading old journal articles and typing in the sequences. The NIH, reflecting this latter view, attempted to create a database with this simple model in mind. For many, the data bank was a "service" and therefore dubiously worthy of federal support under the aegis of basic research. Those at the NIGMS who supported the project had to work hard to generate financial support by stressing the wide range of researchers, including academic, industrial, and medical, who would use the database for basic research.[43] Moreover, Jordan and her co-workers promised that the intention of the funding was only to effect a "start-up" and that it was anticipated that the database would ultimately be supported by user charges.[44] Like lab apparatus or journal subscriptions, the biological database was understood to be something that researchers could pay for out of their own budgets. While providing support for basic researchers, it was not an activity that would contribute fundamentally to biological understanding.

The NIH issued a request for proposals for a nucleic acid sequence database on December 1, 1981. Three proposals were forthcoming: one from Dayhoff and the NBRF, one based on a collaboration between Los Alamos and IntelliGenetics (a company based in Palo Alto, California, and run by Stanford biologists and computer scientists), and a further joint proposal between Los Alamos and Bolt, Beranek and Newman (BBN) of Cambridge, Massachusetts.[45] On June 30, 1982, the NIGMS announced that a contract of $3.2 million (over five years) had been awarded to BBN and Los Alamos. Los Alamos was to be responsible for collecting sequences from the published record, while BBN was to use its expertise in computation to translate the data into a format suitable for distribution by magnetic tape and over dial-up connections to the PROPHET computer (an NIH-funded machine based at BBN). The NBRF was especially disappointed by this decision; others in the com-

munity, too, were concerned about the choice of a nonacademic institution to manage the distribution efforts.[46]

Despite the fact that Goad's pilot project had used a sophisticated
database structure, the NIH insisted that the new data bank—which
would become GenBank—be built as a "flat file." A flat file is a text-
based computer file that simply lists information about nucleotide sequences line by line. Each line begins with a two-letter code specifying
the information to be found on that line—"ID" gives identifying information about the sequence, "DT" gives the date of its publication,
"KW" provides keywords, "FT" lists features in the sequence, and the
sequence itself corresponds to lines beginning with "SQ." Different sequences could be listed one after another in a long text file separated by
the delimiter "//" (figure 5.1).

The NIH held the view that the GenBank format should be readable both by computers and by humans. By using the two-letter line
identifiers, a simple program could extract information from the flat-file
entries. A major disadvantage of this format, however, was the diffi-
culty involved in updating it. If, for instance, it was decided that it was
important to add a further line including information about the type
of sequencing experiment used to generate the sequence, the database
curators would have to modify each sequence entry one by one. Moreover, a flat file does not lend itself to the representation of relationships
between different entries—the list format makes it impossible to group
entries in more than one way or to link information across more than
one entry.

The flat-file format was suited to the NIH's notion that a nucleotide
database should be no more than a simple collection, a laundry list of
sequences. However, it also embodied a particular way of understanding
biology and the function of genes. George Beadle and Edward Tatum's
"one gene–one enzyme" hypothesis is considered one of the founding
dogmas of molecular biology. Although the idea (and its successor, "one
gene–one polypeptide") had been shown to be an oversimplification
even by the 1950s, the notion that it is possible to understand life by
considering the actions of individual genes exerted a profound influence
on at least forty years of biological research.[47]

In the late 1970s, as a result of the sequencing methods invented by
Allan Maxam, Walter Gilbert, and Frederick Sanger, the possibility of
discovering the mechanism of action of particular genes seemed within
reach. Some molecular geneticists began to focus their efforts on finding
and sequencing the genes responsible for particular diseases, such as

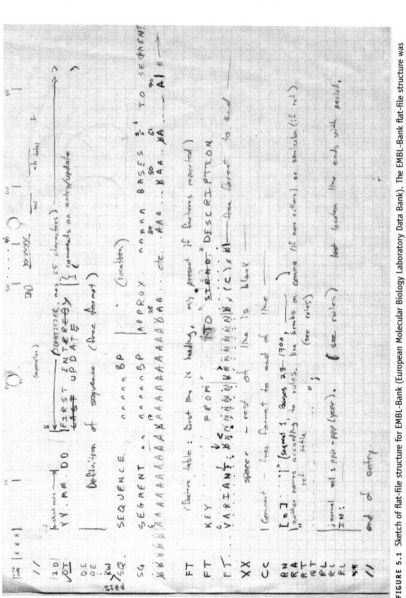

FIGURE 5.1 Sketch of flat-file structure for EMBL-Bank (European Molecular Biology Laboratory Data Bank). The EMBL-Bank flat-file structure was very similar to the one used for GenBank. Note the capital letters on the left ("SQ," "ST," "FT") that define the content of the various lines of the entry. The parts of the entry are listed on consecutive lines, and different entries are simply written into the file one after another, separated by "//". (Greg Hamm personal files. Reproduced with permission.)

cystic fibrosis or Duchenne muscular dystrophy.[48] For such an activity, a flat-file database was an appropriate resource: each sequence entry in the database could be roughly equated with one gene, and that gene had a specific, definite, and singular effect on the biology of the organism in which it dwelt. The NIH imagined that researchers would use the database primarily as a central repository or archive into which sequences would be deposited once and for all; for the most part, sequences would only need to be retrieved one at a time in order to make comparisons with experimental work.[49] Similar sequences usually resulted in similar protein structures with similar functions; hence, matching an unknown sequence to a known sequence in the database could provide invaluable information. For such activities, the gene of interest could be simply compared with the long list of entries in the database one by one. The flat-file database structure was ideal for this kind of search operation; it entailed and represented a theory of how sequence elements acted to produce biological effects.

Once GenBank began operations in July 1982, it became clear to those doing the work of collection and distribution at Los Alamos and BBN that the database was attracting a far wider scope of use. As well as revealing the sequences of genes, the new sequencing technologies had an unexpected consequence: they allowed biologists to sequence not only individual genes, but also regulatory regions, structural RNA-coding regions, regions of unknown function, and even whole genomes (at first limited to small genomes such as those of viruses or cloning vectors). This meant that a sequence in the database did not necessarily correspond neatly to a single gene. Molecular geneticists began to realize that not all the information necessary to understand gene action was contained within the gene sequence—how the gene was spliced, where it was expressed, and how it was phosphorylated were also crucially important.[50] In the flat-file format, such information was contained within the "Features" table for each entry. The Features table consisted of a single line for each feature, as in this example:

| FT | firstexon | EXON | 273–286 |
| FT | tatabox | TATA | 577–595 |

The three columns identified the name of the feature (such as "firstexon"), the feature type (here an exon or a TATA box), and the coordinates in the sequence at which that feature was to be found.[51] Entering this information from the published literature (it was often necessary to read an entire article or even several articles) and keeping it up to date was a gigantic task. But biologists often wanted to use the database to retrieve and aggregate information located across many entries. For instance, a biologist might want to find all the protein-coding sequences in the database that contained exons with a size greater than 100 kilobases. An excerpt from a long list of criticisms of GenBank reads:

> The BB&N [GenBank] retrieval system is not suited to this scientific area. Modern systems permit the user to construct current lists of entries retrieved on various criteria and to perform manipulations on these sequences. The organization of the BB&N system is archaic, because it does not readily permit these manipulations.[52]

The flat file and features table were not well adapted to sophisticated cross-entry queries. Moreover, as biologists produced more and more sequence, it was inevitable that sequences began to overlap; in order for this work to be useful, the database had to identify such overlaps and organize the data in a way that represented these fragments. Another user wrote to Los Alamos complaining that the flat-file data format was not always consistent enough to be computer readable and suggesting "a language for reliably referring to sections of other entries in the database. If this language is sufficiently powerful, many of the synthetic sequences could be expressed in this form."[53] In other words, the user wanted the database to be organized so as to allow the linkages between different entries and different sequences to be made manifest.

The result of these demands was that GenBank was unable to keep pace with the publication of sequences, and particularly with the kinds of annotations that were supposed to appear in the Features table. By 1985, it took an average of ten months for a published sequence to appear in the database. This was not only an unacceptably long delay from the point of view of researchers, but also stood in breach of GenBank's contract with the NIH (which required sequences to be available within three months). A progress report from early 1985 explained the problem:

> Since the inception of GenBank . . . there has been a rapid increase in both the rate at which sequence data is reported and in the complexity of related information that needs to be annotated. As should be expected, many reported sequences repeat,

correct, extend, or otherwise relate to previous work, and as a result a substantial number—in fact, a majority—of database entries have to be updated each year; thus GenBank is not an archival operation such that an entry, once made, just stays in place.[54]

By this time, some members of GenBank's scientific advisory panel considered the growing backlog an "emergency."[55] Los Alamos responded by requesting more money to employ more "curators" to enter data from the published literature. Goad and his co-workers, however, realized that the root of the problem was that the structure of the database was increasingly inadequate for the needs of biological research. James Fickett and Christian Burks, leading contributors to the Los Alamos effort, argued that "the scope and interconnectedness of the data will grow at a pace hard to keep abreast of," and that consequently, the greatest challenge would be to "organize the data in a connected way."[56]

Because the NIH saw the nucleotide sequence database as a mere archiving activity, they attempted to create an atheoretical database. This was impossible: even the minimalist flat file encoded a particular structure, a particular way of doing biology, and a particular idea about how sequences related to organismic function. The flat-file structure instantiated an ordering of biological elements based on the one gene–one enzyme hypothesis. During the early 1980s, that hypothesis was in the process of being displaced and superseded by other ideas about how biology worked.

Biological Relations

The original GenBank contract ran for five years, expiring in September 1987. As that date approached, two concerns were paramount. First, GenBank continued to struggle to remain up to date in entering sequence information from journals.[57] Second, it was clear that the structure of the database required a significant overhaul. As such, NIH's new request for proposals specified that the contractor for the next five-year period would develop a new system whereby authors would be able to submit their sequence data directly in electronic form (preferably over a dial-up telephone network). In addition, the contractor would be obligated to find ways to increase the cross-referencing of the data and to make sure that "new data items which become important can be added to the data base without restructuring."[58] The NIH received three "competitive" proposals for the new contract: one from BBN, one from DNAStar (a company based in Madison, Wisconsin), and one from IntelliGenetics. Each of the contractors would subcontract with Los Alamos. Of singular importance in the eventual decision to award the contract to IntelliGenetics was the perception that it, more than BBN, was in touch with the needs of the biological research community. IntelliGenetics had close ties to the molecular biologists at Stanford—particularly Douglas Brutlag—and had successfully run BIONET, a network resource for providing software tools for biologists, since 1983.[59] No doubt the NIH hoped that a greater awareness of the research needs of molecular biologists would translate into a more usable and flexible database system.

At around this time, many biologists were beginning to think about biology in new ways. The first plans for determining the sequence of the entire human genome were made at a meeting in Santa Fe, New Mexico, in 1986.[60] Even at this early stage, the planners of what came to be called the Human Genome Project (HGP) realized the need for "computational technology" capable of "acquiring, storing, retrieving, and analyzing" the sequence data.[61] Since both Los Alamos and the early stages of the HGP were funded and organized by the Department of Energy, GenBank personnel were well aware of the plans for a massive scaling up of sequencing efforts and the effect that it could have on their already strained ability to get data into the database in a timely fashion. Those advocating the HGP were soon talking to Goad and other GenBank staff about the demands that their project would place on GenBank. By 1988, James Watson, in his capacity as director of the National Center for Human Genome Research (NCHGR), was well aware of the importance of GenBank for the HGP:

> Primary products of the human genome project will be information—genetic linkage maps, cytological maps, physical maps, DNA sequences. This information will be collected and stored in databases, from which it will be made available to scientists and clinicians. In this sense, the raison d'etre of the genome project is the production of databases.[62]

Los Alamos and IntelliGenetics too realized that data coming from the HGP would not only strain the capacity of their staff, but also require thoroughgoing structural changes. In 1985, the complete sequence of the Epstein-Barr virus (about 170,000 bases) had already caused trouble for BBN's computers.[63] In 1988, a "technical overview" of GenBank reported that the addition of human genomic data would require the database to "store entirely new types of data that could not be easily integrated into the original structure."[64] As plans for the HGP (and other smaller genome projects) were developed, the concept of what a sequence database was, and what it could be used for, had to be rethought.

The flat-file database, much like the early file management systems, created a rigid ordering of entries with no explicit cross-linking possible. A relational model would impose different kinds of orderings on the data. The 1988 technical overview of GenBank justified the change to a relational model on the following bases:

> One, because the domain of knowledge we are dealing with is extremely dynamic at this point in history, we had to expect our understanding of the data to change radically during the lifetime of the database. The relational model is well suited to such applications. Two, even if our view of the inherent structure of the data did not change, the ways in which the data could be used almost certainly would change. This makes the ease of performing ad hoc queries extremely important.[65]

By the end of 1986, GenBank staff at Los Alamos had worked out a structure to implement GenBank in relational form. Their plan was set out in a document titled "A Relational Architecture for a Nucleotide Sequence Database," written by Michael Cinkosky and James Fickett.[66] The schema included thirty-three tables that described the sequence itself, its physical context (for instance, its taxonomy or the type of molecule it represented), its logical context (features such as exons, genes, promoters), its citations, and pertinent operational data (tables of synonyms). Tables could be modified or added to (or extra tables could even be added) without disrupting the overall structure or having to amend each entry individually.

The descriptions of the "sequences" and "alignments" tables are reproduced here. Each sequence is given an accession number that acts as the primary key for the table. The "publication_#" and "reference_#" keys link to a table of publications, and "entered_by" and "revised_by" keys link to tables of people (curators or authors). As is noted in the description, such sequences may not correspond to actual physical fragments—that is, they may not represent a particular gene or a particular sequence produced in a sequencing reaction. Rather, the relationship between sequences and physical fragments is "many-to-many": a fragment may be made up of many sequences, and any given sequence may be a part of multiple fragments. In other words, there is no straightforward relationship between DNA sequences as they appear in the database and objects such as "genes" or "exons" or "BAC clones."

```
TABLE sequences
UNIQUE KEY (sequence_#)
INDEX KEYS (publication_#, reference_#), (entered_by, entered_date)
sequence_#     REQ     /* accession number for the sequence */
sequence       REQ     /* the sequence itself */
length         REQ     /* redundant, but convenient */
topology       OPT     /* circular, linear, tandem, NULL-unknown */
publication_#  OPT     /* next two give bibliographic source */
reference_#    OPT
entered_date   OPT     /* next two give history of initial entry */
entered_by     OPT
revised_date   OPT     /* next two give history of revision */
revised_by     OPT
```

DESCRIPTION. The reported sequences. There can be at most one citation, so it is given here. But the relationship to physical fragments can be many-many, so that is given in a separate table.

```
TABLE alignments
UNIQUE KEY (alignment_#, sequence_1, left_end_1, sequence_2)
alignment_#    REQ     /* accession number for alignment */
sequence_1     REQ     /* next three specify first interval to align */
left_end_1     REQ
right_end_1    REQ
sequence_2     REQ     /* next three specify second interval to align */
left_end_2     REQ
right_end_2    REQ
preference     OPT     /* 1 or 2; which one to prefer */
type           OPT     /* conflict, revision, allele, etc. */
```

DESCRIPTION. Give an alignment of any number of sequences by specifying pairs of intervals for the line-up. One record of this table gives a pair of intervals, one from each of two sequences. The set of all records with a given alignment number gives a complete alignment.

This structure for storing sequence data allows objects of interest to be reconstructed from the sequences in multiple ways as needed. The second table shown here—"alignments"—allows different entries in the "sequences" table to be stitched together in multiple ways by referring to their sequence accession numbers and coordinates. For example, it would be possible to create an alignment that spliced sequence A to sequence B, or the first 252 base pairs of sequence A to the last 1,095 base pairs of sequence B. With sufficiently sophisticated queries, it would be possible to join not only sequences, but also any features described in the tables (for example, to join all the exons from given sequences to reproduce a protein-coding region). Sequence data could be linked together dynamically by the user in a flexible manner. But within this flexibility, this relational structure emphasizes the rearrangement of sequence elements. If the flat-file structure was gene-centric, the relational database was alignment-centric. It was designed to make visible the multiple possible orderings, combinations, and contexts of sequence elements.

By 1989, over 80% of GenBank's data had been imported into the relational database.[67] The HGP and the relational sequence database could not have existed without each other—they came into being together. GenBank and the HGP became mutually constitutive projects, making each other thinkable and doable enterprises. Moreover, just as flat files had, both genome projects and relational database systems embodied a particular notion of biological action: namely, one centered on the genome as a densely networked and highly interconnected object. In 1991, when Walter Gilbert wrote of a "paradigm shift" in biology, he argued that soon, "all the 'genes' will be known (in the sense of being resident in databases available electronically), and that the starting point of a biological investigation will be theoretical."[68] This "theory" was built into the structure of the database: phenotype or function does not depend on a single sequence, but rather depends in complicated ways on arrangements of sets of different sequences. The relational database was designed to represent such arrangements.

During the 1990s, biologists investigated the "added value that is provided by completely sequenced genomes in function prediction."[69] As the complete genomes of bacterial organisms, including *Haemophilus influenzae*, *Mycoplasma genitalium*, *Methanococcus jannaschii*, and *Mycoplasma pneumoniae*, became available in GenBank, biologists attempted to learn about biological function through comparative analysis. The existence of orthologs, the relative placement of genes in the genome, and the absence of genes provided important insights into the relationship between genotype and phenotype.[70] The important differences among the bacteria and how they worked were not dependent on individual genes, but on their arrangements and combinations within their whole genomes. But this was exactly what the relational structure of GenBank was designed to expose—not the details of any particular sequence, but the ways in which sequences could be arranged and combined into different "alignments."

GenBank as a relational database provided a structure for thinking

about biology through the genome. It made possible orderings and re-
orderings of biological elements and reinforced biologists' notion that
function depends on multiple sequence elements acting together in interconnected ways.

NCBI And Biological Databases in the Genomic Age

In his opening remarks at the celebratory conference marking the
twenty-fifth anniversary of GenBank in 2008, Donald Lindberg remem-
bered the transformative effect of a paper published in *Science* by Re-
nato Dulbecco. Dulbecco argued that sequencing the human genome
would be a national effort comparable to the "conquest of space." This
argument convinced Lindberg, who was the director of the National Library of Medicine (NLM) at the NIH, that the genome project had to be
undertaken and that the NLM should play a key role. This commitment
was reflected in the NLM's "Long Range Plan" for 1987:

> Currently no organization is taking the leadership to promote
> keys and standards by which the information from the related
> research data bases can be systematically interlinked or retrieved
> by investigators. The full potential of the rapidly expanding in-
> formation base of molecular biology will be realized only if an
> organization with a public mandate such as the Library's takes
> the lead to coordinate and link related research data bases.[71]

During 1986 and 1987, Lindberg worked to convince Congress of the
importance of this mission. The campaign was taken up first by Representative Claude Pepper (D-Florida), who introduced the National
Center for Biotechnology Information Act of 1986. This bill would give
the NLM the responsibility to "develop new communications tools and
serve as a repository and as a center for the distribution of molecular
biology information" (H.R. 99–5271). The NLM circulated a document
on Capitol Hill, titled "Talking One Genetic Language: The Need for a
National Biotechnology Information Center," that made the case for the
new center.[72] Pepper reintroduced the bill with minor modifications in the
next session of Congress (H.R. 100–393), while Senator Lawton Chiles
introduced similar legislation into the Senate on June 11, 1987 (S. 100–
1354).[73] The bill entered Congress at the same time the debates about
the HGP were taking place (Senator Pete Dominici [R-New Mexico] introduced legislation to fund the HGP on July 21). The bill was amended
once more and introduced a third time by Senators Chiles, Dominici,

Ted Kennedy, and others in December 1987 (S. 100–1966). Hearings were held on February 22, 1988, at which Victor McKusick, James Wyngaarden (director of the NIH), and Lindberg testified. Supporters of the bill had closely connected it to the HGP, portraying the need for biotechnology information coordination as central to the project and important for American competitiveness in biotechnology. As support for the HGP grew, the bill's passage became more likely; it was signed into law by President Reagan on November 4, 1988. It provided for the creation of a National Center for Biotechnology Information (NCBI), under the auspices of the NLM, with twelve full-time employees and a budget of $10 million per year for fiscal years 1988 through 1992.[74]

Lindberg conceived the role of the NCBI not as a replacement or supplement for GenBank, but as a way to bring order to the different kinds of biological information and databases that had begun to proliferate. In his testimony in support of the legislation, Donald Fredrickson, president of the Howard Hughes Medical Institute, argued that the NCBI was necessitated by the fact that "not only are the databases being flooded with information they cannot manage, but each database uses a different information system or computer language. We have created a sort of Tower of Babel."[75] "Talking one genetic language" characterizes how the NCBI sought to coordinate diverse sorts of biological information from many sources and at many levels, from cell types to pharmaceuticals. By the time funds for the NCBI were appropriated, Lindberg had already recruited David Lipman to direct the new center. Lipman had been working in Bethesda since 1983 and was already widely respected in the small community of computational biologists for his contribution to sequence-matching algorithms. In the existing biological databases, Lipman saw a tangled mess of overlapping systems and overly complicated schemas; he brought a youthful energy to the task of integrating databases and restoring sense and simplicity to GenBank and other biological information resources.[76] Under Lipman's direction, the NCBI moved quickly to take over GenBank, arguing that its mission to integrate and link databases required close control.[77] By October 1989, it had been agreed that after the end of the current Gen-Bank contract, control of the database would be passed from NIGMS to NCBI—it would be managed in-house rather than under contract to a third party.[78] NCBI took over the task of collecting nucleotide sequence data as Los Alamos' role was phased out.

Before GenBank formally arrived at NCBI in 1992, efforts were already under way to fundamentally change its structure. Commensurate with the overall mission of the NCBI, the aim was the make GenBank data more amenable to integration and federation with other data types. The NCBI data model was the work of James Ostell, who had been asked by Lipman to join NCBI as its chief of information engineering in 1988.[79] Ostell needed to solve two problems. The first was how to make data available to the widest possible number of biological users by ensuring that they could be shared across different computer platforms. Ostell's solution was to adopt an international standard (ISo8824 and ISo8825) called ASN.1 (Abstract Syntax Notation 1). Like the hypertext transfer protocol (HTTP) used on the Internet, ASN is a way for computers to communicate with one another—it specifies rules for describing data objects and the relationships between them. Unlike HTTP, however, it is not text-based, but renders data into binary code. ASN.1 was developed in 1984 for the purpose of structuring email messages; it describes in bits and bytes the layout of messages as they are transmitted between programs or between different computers. ASN.1 acts as a universal grammar that is completely independent of any particular machine architecture or programming language.[80] Ostell chose ASN.1 because "we did not want to tie our data to a particular database technology or a particular programming language."[81] Using ASN.1 meant that biologists using any programming language or computer system could use the GenBank database.

The second problem was to find a way of storing various kinds of data in a form that was suited to the needs of biologists who wanted not just DNA sequence information, but also data about protein sequence, protein structure, and expression, as well as information contained in the published literature. The scale of this problem of "heterogeneous sources" had become such that relational databases were no longer appropriate for such linking. "It is clear that the cost of having to stay current on the details of a large number of relational schemas makes this approach impractical," Ostell argued. "It requires a many-to-many mapping among databases, with all the frailties of that approach."[82] In other words, keeping the structure of each database consistent with the structure of a large number of others would quickly prove an impossible task. The alternative was to find a way to link the databases using ASN.1 via what Ostell called a "loose federation." The first such application, which became known as Entrez, used ASN.1 to link nucleic acid databases, protein sequence databases, and a large database of biomedical literature (MEDLINE). Wherever an article was cited in a sequence database (for instance, the publication from which the sequence was taken), the NCBI created a link to the relevant article in MEDLINE using the MEDLINE ID (figure 5.2). Likewise, NCBI created links

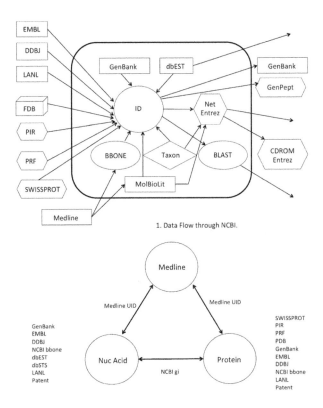

1. Data Flow through NCBI.

2. Entrez is three linked data spaces.

3. Five linked information spaces

FIGURE 5.2 Integration of data at NCBI. Diagrams show the emphasis on connections between different databases such as sequence, protein, MEDLINE (publications), dbEST (expressed sequence tags), and so on. (Ostell, "Integrated access." Reproduced by permission of IEEE.)

between nucleotide sequences and their translated protein equivalents. "Once the work has been done to get this kind of cross-reference into the databases," Ostell wrote, "it becomes relatively straightforward to tie them into a unified system using hypertext tools such as Mosaic."[83] The new structures for organizing biological information were closely connected to the new tools of the Internet.

But even building these kinds of hard links between databases was too difficult. For Ostell, the ultimate solution was to create a "data model." Like animal models, which biologists can use for the study of human diseases, or mathematical models, which they can use for de-scribing forces inside cells, the NCBI data model provides a structure for sequence information that allows "meaningful predictions to be made and tested about the obviously much more complex biological system under consideration."[84] Ostell reasoned that the basic elements of the database should, as closely as possible, resemble the basic "facts" collected by biologists in the laboratory—that is, sequence elements. In the data model, sequences are represented as objects called "Bioseqs," which constitute a "linear, integer coordinate system." Importantly, the sequence information itself (the As, Gs, Ts and Cs) is not contained in the Bioseqs. Rather, a particular Bioseq contains coordinate-based instructions on how to build a sequence from fragmentary pieces of sequenced DNA or RNA. Such instructions could be: "Take *sequence1*, then a gap of unknown length, then *sequence3*, then a gap of 30 base pairs, then *sequence2*." The Bioseq may consist of full-length sequences, partial sequences, gaps, overlapping sequences, or genetic or physical maps, since all of these can be constructed by placing different sorts of objects along the coordinates of the Bioseq. This structure allows for a very different representation of biological data within the system:

> The GenBank flatfile format . . . is simply a particular style of report, one that is more "human-readable" and that ultimately flattens the connected collection of sequences back into the familiar one-sequence, DNA-centered view. [The NCBI data model] much more directly reflects the underlying structure of such data.[85]

Indeed, the aim of the data model was a "natural mapping of how bi-ologists think of sequence relationships and how they annotate these sequences. . . . The model concentrates on fundamental data elements that can be measured in the laboratory, such as the sequence of an iso-lated molecule."[86] This system not only allowed the expression of very complex relationships between sequences and pieces of sequences based on maps or alignments,[87] but also provided sophisticated and robust links to published scientific articles. Citations in various databases were mapped to MEDLINE via unique integer identification numbers. Appropriate software could rapidly search (across multiple databases) for objects that cited the same article and link those objects together, or it could go even further and make links based on keywords from the abstracts contained in MEDLINE. By rendering the model in ASN.1, the NCBI created a system that combined objects (DNA sequences, protein sequences, references, sequence features) from a variety of databases and manipulated them all with a common set of software tools.

DNA-centered relational databases provided more flexible ways to recombine and reorder sequences. ASN.1 and the data model permitted no static biological objects. Rather, it was assumed that the process of doing biology would involve recombination and reordering of different biological objects across a wide range of databases. Relational databases were a framework within which to investigate the properties of dynamically rearrangeable sequence elements. The data model was a framework within which to investigate genomes using a wide variety of other data and data types.

The data model has provided a framework for exemplary experiments of the postgenomic era. Although it was developed in 1990, it remains a powerful tool for moving biological investigation beyond the genome. As biologists began to realize the limitations of studying the genome in isolation, the data model demonstrated ways in which to integrate more and more kinds of biological data.

In 2005, the bioinformatician Hans P. Fischer called for "inventorizing biology"—capturing the entirety of information about an organism in databases. Genomes, transcriptomes, proteomes, metabolomes, interactomes, and phenomes should be characterized, entered into databases, and integrated. This new "quantitative biology" would transform drug discovery and allow us to understand human disease pathways. This vision of "tightly integrated biological data" would allow an engineering-like approach to biological questions—drug design or even understanding a disease would become more like building an aircraft wing.[88] In the postgenomic era, the organization and integration of biological information provides a structure or blueprint from which biologists can work. At the beginning of each year, *Nucleic Acids Research* publishes a "database issue" that provides an inventory of biological databases. In 2009, that list included 1,170 databases, including about 100 new entries.[89] The ways in which the information in those databases is connected provide theories of biological action. It is now clear that the sequence of the genome alone does not determine phenotypic traits. Databases provide ways of linking genomic information to the other vast amounts of experimental data that deal with transcripts, proteins, epigenetics, interactions, microRNAs, and so on; each of those links constitutes a representation of how sequences act and are acted on in vivo to make life work.

Conclusions

Biological databases impose particular limitations on how biological objects can be related to one another. In other words, the structure of a database predetermines the sorts of biological relationships that can be "discovered." To use the language of Bowker and Star, the database "torques," or twists, objects into particular conformations with respect to one another.[90] The creation of a database generates a particular and rigid structure of relationships between biological objects, and these relationships guide biologists in thinking about how living systems work. The evolution of GenBank from flat-file to relational to federated database paralleled biologists' moves from gene-centric to alignment-centric to multielement views of biological action. Of course, it was always possible to use a flat-file database to link sequence elements or to join protein interaction data to a relational database, but the specific structures and orderings of these database types emphasized particular kinds of relationships, made them visible and tractable.

One corollary of this argument is that biological databases are a form of theoretical biology. Theoretical biology has had a fraught history. In the twentieth and twenty-first centuries, several attempts have been made to reinvent biology as a theoretical, and in particular a mathematical, science. The work of D'Arcy Thompson, C. H. Waddington, Nicolas Rashevsky, and Ludwig von Bertalanffy has stood out for historians.[91] The work of all these authors could be understood as an attempt to discover some general or underlying biological principles from which the facts of biology (or the conditions of life) might be derived and deduced. In the twentieth century, such efforts were almost completely overshadowed by the successes of experimental biology, and molecular biology in particular. As Evelyn Fox Keller recognizes, however, the increasing use of computers in biological research has relied on modes of practice that might be called theoretical. "In molecular analyses of molecular genetics," Keller argues, "observed effects are given meaning through the construction of provisional (and often quite elaborate) models formulated to integrate new data with previous observations from related experiments. As the observations become more complex, so too do the models the biologists must construct to make sense of their data. And as the models become more complex, the computer becomes an increasingly indispensable partner in their representation, analysis, and interpretation."[92] This description might apply equally well to biological databases as to the type of computer models that Keller describes. The structures and categories that databases impose are models for integrating and making sense of large sets of data. As categorizations of organisms, sequences, genes, transposable elements, exon junctions, and so forth, databases are built on sets of structures or principles of biological organization that are then tested in experiments. Far from being lists or collections of information, biological databases entail testable theories of how biological entities function and fit together.

This understanding of biological databases as models also demonstrates that the flow and ordering of data are central to the constitution of biological objects and knowledge in bioinformatics. Here we have once again followed the data into the structures and spaces inside computers. Databases, which summarize, integrate, and synthesize vast amounts of heterogeneous information, are the key tools that allow biologists to ask questions that pertain to large numbers of sequences, genes, organisms, species, and so on. Databases allow these objects to be constituted "out of sequence"—that is, brought into new orderings or relationships with one another that do not necessarily reflect their order in cells or on chromosomes. The form of such relationships is constrained, however—flat files and relational databases were not designed for biology, but rather have their own particular histories. The ways in which biological objects are related to one another have been conditioned by the structural possibilities and limitations of existing database models—that is, by the histories of databases themselves.

6 Seeing Genomes

Chapter 5 showed how biological objects are organized inside computers. But it is no more possible to observe data in a database directly than it is to observe molecules inside a cell directly. So how are data objects actually represented and manipulated by biologists in their work? All the examples in this chapter will show what biologists actually *do* while sitting in front of their computers: they manipulate visual representations of biological objects. These representations act to generate new and often unexpected relationships between biological objects that are crucial in the making of bioinformatic knowledge.

In this chapter we will follow the data out of databases and into the visual realm.[1] Visualizations are not produced as the end results of biological work, or as afterthoughts prepared for publication or for presentations, but rather form an integral part of how computational biologists think about their own work and communicate it to their collaborators. Arranging data into images forms a crucial part of knowledge production in contemporary biology—it is often through visualizations that data are made into knowledge. A large part of bioinformatics is about solving problems of representation. Like the structure of databases, the choice of visual representations has a determinative effect on the knowledge and objects produced through them. Computational representations of biological objects involve decisions about what features of an object are represented and how they are related to one another.

When an object is represented in an image, it is poured into a carefully contrived structure. As with databases, decisions about visualization techniques and tools are decisions about how to constitute an object. These decisions, in turn, affect how the biological object itself can be understood. Bioinformatic "pictures" do not attempt to capture what life looks like, but rather to figure out how it works. The most important visuals, the ones that bioinformaticians use every day, are not like images from a microscope, or even like particle tracks computationally reconstructed from a spark chamber, which are both traces of physical entities. Rather, they break traces into pieces, reorder them with computations, and reconstruct them into pictures that are easily interpretable and full of biological meaning. These pictures are usually the starting points, rather than the end results, of bioinformatic work. These pictures are what biology looks like to many bioinformaticians—they are the raw material with which experiments can be performed. Visualization in bioinformatics is—through and through—analysis and quantification.[2]

As the amount of biological data, especially sequence data, has increased, the usefulness of thinking about biology merely as a long series of AGTC words has diminished. In 2012, GenBank contained around 150 billion nucleotides—this is an inconceivably long "book of life" (more than fifty thousand times the length of the Bible). Ways of rendering data into images, rather than text, have gained in importance. Doing things with large amounts of biological data requires thinking of them as a picture, not merely as a text. The transition from text to image is an ongoing process, but historians and anthropologists of biology should be attentive to the new metaphors—visual and pictorial ones—that will increasingly dominate the discipline.[3]

The Ensembl database, at the European Bioinformatics Institute, is a sophisticated example of a "genome browser," an online tool that biologists can use to see and study genomes. This representation of the genome is almost entirely a visual one; through it, the user-biologist "sees" the genome as a set of color-coded horizontal lines. Of course, this representation is not intended to be a faithful depiction of the intricate, tangled-up set of molecules of a real genome. But in an important sense, this is what the genome is—this is how it looks and feels—to many biologists; visualization techniques become the only way of seeing the object in question.[4] Large matrices or grids are also important for thinking about problems in bioinformatics. Reducing multidimensional data to a two-dimensional array provides both a powerful computational tool and a cartographic metaphor for relating to sets of data. In a biology of massive data sets, high throughput, and statistics, it is the visual that is coming to dominate the textual. Sequence is no longer just script. Computer-based visualization has made (and is making) possible the comprehension of vaster and more heterogeneous data sets. It is therefore helping to justify the massive production of data that characterizes contemporary bioinformatics.

The importance of computer graphics to biology is largely a recent phenomenon. Before the 1980s, most computers were simply not powerful enough to perform image processing. Even before computers could be used as visualization tools, however, biologists deployed visual metaphors for solving data-intensive problems. The first part of this chapter describes how maps and matrices became ubiquitous features of data-driven biology in the 1960s and 1970s. The increasing availability of graphical workstations and personal computers in the 1980s allowed biologists to develop more sophisticated representational tools. The second part describes the development of AceDB for the representation of the worm genome and its evolution into today's genome browsers. The last part concentrates on how visualization tools such as genome browsers and heat maps are actually used in the day-to-day work of biology. These visualization tools are central to how biologists see, manipulate, and know biological objects. The forms these images take are products of the histories of computers, algorithms, and databases.

From MAC to Mac: How Computers Became Visualization Tools

In 1966, Cyrus Levinthal began to use computers to visualize protein molecules. This effort was part of the Advanced Project Research Agency's Project MAC at the Massachusetts Institute of Technology.[5] Levinthal and his co-workers saw their work as a way to replace the ungainly and bulky space-filling physical models that were used to represent proteins. The aim was to allow viewing and manipulation of large macromolecules without the hassle of actually having to build a three-dimensional representation from wood, metal, or plastic. Levinthal attempted to achieve this by designing an elaborate set of systems for interaction between the user and the model on the computer screen: a "light-pen" and a trackball-like device allowed (to quote Levinthal's paper) "direct communication with the display."[6]

As Eric Francoeur and Jérôme Segal have argued, these protein pictures provided a middle way between physical models and purely nu-

merical models of macromolecules.[7] Atomic positions within the proteins were stored as vectors within the computer, so that (to quote from
Levinthal's paper again) "the illusion of rotation and, consequently a
three-dimensionality, can be produced by presenting the observer with a
sequence of projections of an incrementally rotated vector model." The
visual display and the means of manipulating it provided protein modelers with a means of "seeing" numbers stored within the computer in
a way that had biological meaning. Indeed, the main use of the models
was, as Levinthal put it, to "assist in the process of user directed modification of structure." In other words, users would manually alter the
configuration of the protein on their screen in order to refine the structure or to simulate protein folding. The computer did not automatically
calculate the optimum configuration of the protein, but rather allowed
the user to experimentally tweak the model until a satisfactory representation was found. Levinthal's work was an attempt not just to represent
biological macromolecules inside computers, but to allow biologists to
do something with the representations in order to learn about the function and behavior of proteins.

Despite Levinthal's efforts with protein structures in the 1960s,
computer-based visualization in biology did not catch on. As Francoeur
and Segal argue, the advantage of Levinthal's technology over ordinary
physical modeling were not immediately obvious to biochemists, who
had their own research styles (Levinthal himself was a nuclear physicist
by training, not a biologist), and the cost of setting up the computing
equipment was prohibitively high. It is also significant that Levinthal
and his colleagues reported their work not to a conference of biologists
or biochemists, but to a group of computer graphics specialists. Reports
on the use of computers in biomedicine in the 1960s, such as the Air
Force–commissioned *Use of Computers in Biology and Medicine* from
1965, saw the main prospects of computing in data storage and quantitation, mentioning imaging hardly at all.[8] This was no doubt in large
part because most computers of the 1960s were not powerful enough
to process images at all—many were still working by means of punched
cards.

As biologists began to deal with larger and larger amounts of data,
"visual" thinking proved increasingly useful, even if it was not directly
associated with computing. Margaret Dayhoff's work on protein sequences, discussed in detail in chapter 5, was organized around the concept of an "atlas."[9] Dayhoff's *Atlas* looks nothing like an atlas of maps,
nor much like the anatomical or botanical atlases of the eighteenth and
nineteenth century. However, it has in common with other atlases the

aim of offering a complete and organized catalog of information on a particular subject. Just as an anatomical atlas was supposed to summarize knowledge about the form of the human body, the individual pages filled with protein sequence data formed a map of biological knowledge. For Dayhoff, the distinct sequences were maps of a small part of the space of biochemical structure, function, and origins; taken together, they formed a total image, or atlas, of biology. The visual metaphor provided both a rationale and a set of organizing principles for Dayhoff's collecting.

If organizing sequences was difficult, comparing them was even harder. In the 1970s, sequence comparison became an exemplary problem for bioinformatics, since "matching" of sequences provided the key to understanding evolution, protein function, and, later, whole genomes.[10] But the comparison of even relatively short sequences was a tedious task that involved many steps. Many of the methods devised to attack this problem were based on visual metaphors and insights.

The basic method of comparing two protein sequences was to write each sequence on a strip of paper and slide the pieces of paper past one another, one amino acid at a time, to produce different alignments; for each alignment, the biologist would have to count the number of matches and mismatches in order to "score" the alignment. The alignment with the best score (the fewest mismatches) was the best alignment. One alternative, known as the dot-matrix method, was to write one sequence across the top of a matrix and the other down the left side; the biologist then filled in the matrix, placing a dot wherever two amino acids matched. The best alignment could be found simply by inspection: the longest diagonal line of dots was the best alignment (figure 6.1).[11]

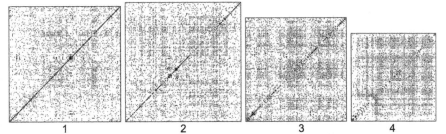

FIGURE 6.1 The dot matrix method of matching sequences. "Dot plots" are constructed by writing a sequence from one organism vertically on the left and a sequence from another organism horizontally across the top. The actual letters of the sequences are not shown in these images. Dots are then placed where the nucleotide in a column matches the nucleotide in a row. Thick diagonal lines appear where there are regions of similarity between two sequences. (Wright et al. "A draft annotation." Reproduced from *Genome Biology* under the BioMed Central open access license agreement.)

Summarizing data in matrix form provided a visual solution to a data analysis problem. In 1970, Needleman and Wunsch developed a more sophisticated matrix-based method based on dynamic programming techniques.[12] This ubiquitous algorithm is almost always explained, and understood, visually: a way of finding an optimum path through a grid and then "tracing back" to find the best alignment of the sequences.[13] In pictorial terms, the algorithm lines up sequences on a grid or matrix—much like the dot matrix method—and then scores various possible matches between nucleotides; the final alignment is found by drawing a line through the boxes with the highest scores.[14] Dynamic programming is not just a way of picturing a protein sequence, but also of analyzing its relationship to another sequence by seeing the alignment as a path through a two-dimensional space. These matrix methods were visual solutions to data- and calculation-intensive problems in early bioinformatics.

The late 1970s and early 1980s saw the development of the personal computer. Rather than having to log into central computing facilities, biologists could now run and program computers in their own laboratories. Computers became more powerful and more oriented toward graphical rather than textual display of information. The Apple Macintosh (first available in 1984), with its graphical user interface, was especially important in this respect: biologists became more accustomed to being able to use spreadsheet programs to rapidly create charts and other images to quickly inspect and analyze their data. Moreover, as more and more biologists connected to various telephonic and electronic networks, the potential for sharing and gathering large amounts of biological data grew rapidly. Visualization and exchange of data became understood as the primary uses for computers in biology. A report produced by the National Academy of Sciences in 1989 noted on the first page of the executive summary that "analytic capabilities have improved significantly, along with the capacity to present results as visual images."[15] A report to the National Science Foundation, published in the journal *Computer Graphics* in 1987, saw the overwhelming growth of data as the immediate problem. The solution was visualization: "Scientists need an alternative to numbers. A technical reality today and a cognitive imperative tomorrow are the use of images. The ability of scientists to visualize complex computations and simulations is absolutely essential to ensure the integrity of analyses, to provoke insights, and to communicate those insights with others."[16] Some went so far as to hail a "second computer revolution," in which "the ultimate impact of visual computing will eventually match or exceed the tremendous societal changes already wrought by computers . . . computers with a window open up a whole new kind of communication between man and machine."[17] This last was perhaps an extreme view, but the late 1980s and early 1990s saw the creation and use of a range of new techniques that used visualization not just as a mode of communication, but as a method of calculation and analysis.

Already in the 1960s, those using computers to do biological work were using visual metaphors and pictorial means to arrange and understand their data. As I argued in chapter 5, Dayhoff's aim was the organization of an otherwise vast and unordered glut of information about protein sequence; the *Atlas* helped her achieve this end. Comparison of sequences can also be understood as a problem of the proper ordering and arrangement of data (the set of all amino acid to amino acid comparisons); this too could be understood as a visual problem of "alignment." These spatial organizations provided ways of ordering biological objects that suggested meaningful connections between them. They were not supposed to faithfully depict sequence or structure, and they were not directed at fabrication, but rather at depicting new and unexpected relationships between biological objects.

From AceDB to the Genome Browsers

By the early 1990s, genomic data were beginning to proliferate, and biologists worried about their ability to synthesize and analyze all this sequence. In 1989, Richard Durbin, working at the Laboratory of Molecular Biology at Cambridge, UK, and Jean Thierry-Mieg, from the Centre National de la Recherche Scientifique (CNRS) at Montpellier, France, were trying to find a way to work with data from the *C. elegans* genome project. They required a system that would both maintain the data and make them accessible to the *C. elegans* research community. The physical clone map for *C. elegans* had been compiled using a program called CONTIG9; this program was distributed, together with the map data, as a program called PMAP.[18] Durbin and Thierry-Mieg built "A *C. elegans* Data Base" (AceDB), as it came to be known, as an extension of PMAP: the idea was to store sequence, physical and genetic maps, and references within a single database. To build in the necessary flexibility, the authors quickly made AceDB into more than a database. First, it became a general database management system. "Since this is a new type of system," the authors reasoned, "it seems very desirable to have a database whose structure can evolve as experience is gained. However, this is in general very difficult with existing database systems, both relational, such as SyBase and Oracle, and object-oriented, such as ObjectStore."[19] It was not only the quantity of the new genomic data, but also the need for integration of different types of data that would be required to understand them, that motivated Durbin and Thierry-Mieg's work. A database was necessary not merely as a storage mechanism, but also as a way to help the brain of the biologist to organize and synthesize data into forms that could be used to answer biological questions.[20] It was a tool for organizing and manipulating data in order to help biologists find interesting relationships within those data. How could this be achieved?

Durbin and Thierry-Mieg's second innovation was making AceDB into a visualization system. Their description is worth quoting at length:

> Overall the program is very graphical. It works using a windowing system, and presents data in different types of window according to the different types of map. The maps and other windows are linked in a hypertext fashion, so that clicking on an object will display further information about that object in the appropriate sort of window. For example, clicking on a chromosome displays its genetic map; clicking on a gene in the genetic map displays text information about the gene's phenotype, references etc.; clicking on a clone displays the physical map around it; clicking on a sequence starts the sequence display facility.[21]

AceDB allowed researchers to navigate around a genome just as they could move through the pages of the new World Wide Web. Figure 6.2 shows a screenshot of AceDB with a genome map displayed vertically down the screen. Durbin and Thierry-Mieg realized that solving biological problems in the coming genomic age meant solving problems of data synthesis by finding appropriate modes of representation and visualization.

The designers of AceDB attempted to minimize the difference between the internal (that is, inside the computer) representation of genomic objects and their visual representations on the screen. To achieve this, they organized the database as a series of "objects," each having a treelike structure; the ends of the branches of each tree pointed either to another object or to the data themselves (numbers or characters). For example, a "Gene" object could include branches that pointed to information about its allele, its clone name, its physical map location, and so

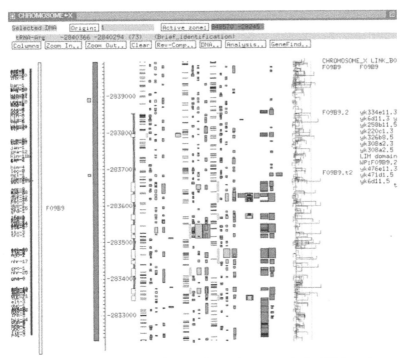

FIGURE 6.2 AceDB vertical description of genome. The area shown represents about 700 base pairs (roughly 283300–284000) on the X chromosome of the worm genome in the region of a gene called F09B9. The various boxes in the middle of the image indicate the extent of various features such as exons, SNPs, and coding start and stop sites. (http://www.acedb.org/Cornell/angis/genetic_maps.html. Reproduced by permission of Jean Thierry-Mieg.)

on. A simple text file format sufficed to define this structure. Each object belonged to a particular "class" that defined the organization and types of data permitted for the object.

This system of classes and objects is familiar to software engineers as "object-oriented" programming. In an object-oriented programming language, classes define the abstract characteristics of objects (both things that an object has and things that it can do), while the objects themselves are particular instances of a class. The advantage of such a scheme is that it is supposed to mirror how we think of objects in the real world.[22] For example, one could define the class "vehicle" as including any object with "wheels" and a "power source"; subclasses would include "car," which must have four "wheels" plus an "engine"; "truck," which must have six "wheels" or more plus an "engine"; and "bicycle," which must have two "wheels" plus two "pedals." The particular car that I own would be an object within the class "car." The first object-oriented programming language, called Simula, was developed at the Norwegian Computing Center beginning in 1963 under a contract with the Sperry Rand Corporation. The need for a language that could be used to simulate complicated systems derived largely from the field of operations research.[23] The problem was to design a language that could easily and accurately describe objects and events in the world:

> Programming cannot be started before the system is precisely described. There is a strong demand for basic concepts useful in understanding and describing all systems studied by simulation from a common point of view, for standardized notation, for easy communication between research workers in different disciplines.[24]

In other words, the key to solving complicated simulation problems lay in representing them inside the computer in a way that matched descriptions in the real world.

Durbin and Thierry-Mieg were faced with a similar problem—that of representing and relating biological objects (some of them yet undiscovered or undescribed) inside a computer in a way that would be recognizable and useful to biologists. Their object-oriented database aimed to provide a means of describing biological objects from the real world. In doing so, however, AceDB also placed constraints on what those objects were. In order to represent the objects, AceDB's designers had to make decisions about what they *were*—for instance, how would the class "Gene" be described? What kind of objects could it contain? How should the hierarchy of classes be structured? These sorts of representational decisions necessarily produced the meaning of genes, genomes, exons, and so forth as highly structured objects inside the database.

For Durbin and Thierry-Mieg, storing, sharing, viewing, and understanding *C. elegans* data were all part of the one problem. AceDB's visualization is something that biologists can not only observe, but also work with, manipulate, and analyze. As such, the ways in which biological objects are structured within the database have consequences for how those objects are understood. Because AceDB provides a way of "seeing" biological objects, the features that it emphasizes become the features of the object that become "visible" and therefore pertinent to biological work. Just as a single territory can be rendered as a political map, a physical map, a topological map, and so on, AceDB produces "maps" of the genome that show how various objects map onto the "territory" of the sequence. Under the gaze of AceDB, this territory—

the sequence itself—becomes a particular sort of space with a particular structure and with particular features. The computational representation has thoroughgoing consequences for how biological objects are understood.

AceDB is a remarkably successful tool. The database quickly spread from its original use as a worm sequencing tool to other whole-organism sequencing projects. The Wellcome Trust Sanger Institute and the Genome Institute at Washington University, both major laboratories responsible for mapping and sequencing the human genome, adopted AceDB. More recently, it has also been adapted for uses with nonbiological data.[25] By the late 1990s, the amounts of biological (and especially sequence) data from a number of species, including humans, had grown so much that it was realized that visual tools such as AceDB were going to be crucial for understanding and managing this volume of information. The problem was that, for biologists, the genome in its raw, textual form was not usually a very useful thing to work with—who can make sense of 3 billion As, Gs, Ts, and Cs? Biologists needed ways of viewing these data, of seeing them in a way that immediately brings their salient features (such as genes) into view. As Durbin and Thierry-Mieg wrote, "Clearly what is required is a database system that, in addition to storing the results of large scale sequencing and mapping projects, allows all sorts of experimental genetic data to be maintained and linked to the maps and sequences in as flexible a way as possible."

As more and more of the human genome was sequenced, the need for a powerful system of annotation became ever more pressing. In 1998, Celera Genomics, a private company, challenged the public HGP, claiming that it could sequence the human genome better and faster. As the rivalry between the public consortium and Celera intensified, the project directors realized that the battle would be lost or won not just though sequencing speed, but also through the presentation and representation of the genomic data.[26] Durbin put one of his graduate students to work on the problem. Ewan Birney, along with Tim Hubbard and Michele Clamp, generated the code for what became Ensembl, a "bioinformatics framework to organize biology around the sequences of large genomes."[27]

Chapter 4 has already described some of my experiences working with the team responsible for maintaining Ensembl at the European Bioinformatics Institute (EBI). I will elaborate on that work here in order to describe how AceDB evolved to cope with the even larger amounts of data that emerged after the completion of the HGP. Ensembl's primary goal was to solve the problems of representing such large data sets in ways that could be understood and used by biologists. For example, the most immediate problem was how to represent huge objects like genomes in a useful way. Although AceDB provided a model for many of their ideas, the EBI team decided to replace its hierarchical file structure with a relational database management system that could provide more adaptability to the large data volumes associated with the human genome.[28]

The basic unit of representation in Ensembl, the "contig," was dictated by the way sequence was produced in high-throughput sequencing machines. A contig represented a contiguous piece of sequence from a single bacterial artificial chromosome (BAC)—the laboratory construct used to amplify and sequence the DNA. The relationship between the contigs and the other information in the database is shown in figure 6.3. The letters of the DNA sequence itself are stored in a table called "dna," while the instructions for assembling a complete chromosome from several contigs are contained in a separate table called "assembly."[29] As the Ensembl developers explain, "a gene from Ensembl's perspective is a set of transcripts that share at least one exon. This is a more limited defini-

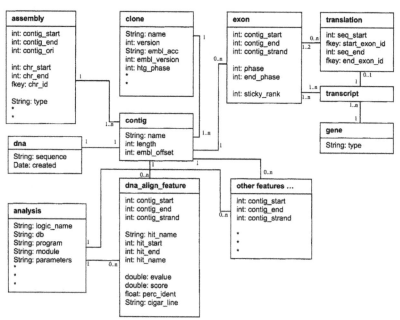

FIGURE 6.3 Ensembl database schema. The "contig" is the central object; each contig corresponds to one DNA sequence (listed in the table "dna sequence") but may have multiple features ("dna_align _feature") and exons ("exon"). Contigs can be put together into multiple assemblies ("assembly") which stitch together various contigs. (Stabenau et al., "Ensembl core software." Reproduced with permission of Cold Spring Harbor Laboratory Press.)

tion than, for example, a genetic locus, but it describes a relationship that can be easily identified computationally."[30] Of course, no laboratory biologist would think of a gene in this way. To bridge this gap, Ensembl's designers created a set of "adapters," or Application Program Interfaces (APIs), that mediate between the database and the user. The APIs are software elements, written in the object-oriented programming language Perl, that create familiar biological objects from the database:

> Ensembl models real-world biological constructs as data objects. For example, Gene objects represent genes, Exon objects represent exons, and RepeatFeature objects represent repetitive regions. Data objects provide a natural, intuitive way to access the wide variety of information that is available in a genome.[31]

The APIs make it possible for software accessing the database to use concepts familiar to laboratory biologists. For example, code to retrieve a slice of DNA sequence from a region of chromosome 1 would be written as follows:

```
#makes a connection to the database
my $db=Bio::EnsEMBL::DBSQL::DBAdaptor → new
    ( -host → 'ensembl.db.org,'
    -user → 'anonymous,'
    -dbname → 'homo_sapiens_core_19_34'
    );
#gets an object (called a slice adaptor)that slices up sequences
my $slice_adaptor=$db→get_SliceAdaptor();
#uses the slice adaptor to retrieve the sequence of chromosome 1 between
coordinate 10 million and 20 million
my $slice = $slice_adaptor→fetch_by_chr_start_end
    ('1,'10_000_000, 20_000_000
    );
#extracts particular features from this piece of sequence
my $features=$slice→get_all_protein_align_features
```

The language of Perl suggests motion and action: slices, adaptors, "get," "fetch," create the sense that the program is moving around and cutting up real objects.

To create a further level of abstraction between the database and the user, Ensembl also allows any piece of sequence to be treated as if it were a single database object. The use of "virtual contigs" means that the biologist can effectively treat the sequence as a continuous object despite its underlying fragmented representation. The Ensembl designers explained the importance of the virtual contig concept for both the computer and the user:

> [The virtual contig object] allows access to genomic sequence and its annotation as if it was a continuous piece of DNA in a 1-N coordinate space, regardless of how it is stored in the database. This is important since it is impractical to store large genome sequences as continuous pieces of DNA, not least because this would mean updating the entire genome entry whenever any single base changed. The VC object handles reading and writing of features and behaves identically regardless of whether the underlying sequence is stored as a single real piece of DNA (a single raw contig) or an assembly of many fragments of DNA (many raw contigs). Because features are always stored at the raw contig level, "virtual contigs" are really virtual and as a result less fragile to sequence assembly changes. It is this feature that allows Ensembl to handle draft genome data in a seamless way and makes it possible to change between different genome assemblies relatively painlessly.[32]

The virtual contig object mediates between the computer and the lab, both in allowing the database to be adaptable to the vagaries and uncertainties of real biological data and in presenting the final sequence in a way that makes sense to the biologist. But virtual contigs are also abstract representations that allow bioinformaticians to cut away the messiness of the internal database representations; in fact, they are highly structured objects whose properties are determined by the Perl classes to which they belong. Once again, the representation (the virtual contig) alters what it is possible to do with—and how the biologist imagines—the biological object itself (the contig as an experimental object of sequencing). Biologists' intuition about the object is based on the properties of its "adaptors"—that is, on how it can be moved around and manipulated inside the database. In mediating between biological and computational representations, these sorts of structures come to redefine biological objects themselves.

The dependence of biology on computational representations becomes even clearer if we examine the final and most frequently manipulated layer of Ensembl: the web interface. It is at this point that biologists interact with genomes every day in their routine work. The images on the screen are not generated directly from the database. Rather, when a remote user navigates to a particular region of the genome, the web server retrieves the relevant information from the database via the API; this information is then used to dynamically generate a web page for the user.[33] The user can view the genome in various ways, either scrolling along an entire chromosome and zooming in and out or as a map of genetic markers, genes, and proteins. The aim of the Ensembl genome browser is to provide easy access to specific information, to provide a high-level overview, or to provide a way to examine patterns in the data. As such, a great deal of attention is paid to the details of the display: the organization of information on the page, the speed at which it loads, and the colors and shapes used to represent various objects were constantly under discussion among the Ensembl team. During the time I visited, the team released an updated version of its website. The design changes they described focus almost exclusively on visual issues:

> One of the continuing challenges with a site that contains so much data is making it discoverable. In testing sessions we found that users were overlooking the left-hand menu with its links to additional data, perhaps because there was so much on the page already that they didn't expect to find even more! . . . The main change we have made is to improve the way that users move between pages. The displays have been grouped into Location, Gene, Transcript and Variation and tabs at the top of the page allow the user to switch easily between these sets of views. If you browse to a location on a genome, initially only the Location tab is present, but the others appear as you select specific genes and transcripts to view. . . . You will also have noticed that we have changed the colour scheme somewhat, darkening the blue background and extending it to the whole page header, and using the yellow background to draw attention to important parts of the page such as genomic images and information boxes.[34]

The redesign was not just an effort to make the site visually appealing: the Ensembl bioinformaticians realized that how they displayed the objects would exert a determinative influence on whether and how particular data elements were used and understood by laboratory biologists. Deciding how to represent biological objects had important consequences for what those objects were and how they could be used.

At roughly the same time as Birney and his team, biologists in the United States also began to develop tools for seeing whole genomes. One of the most successful—the UCSC Genome Browser—was developed at the University of California, Santa Cruz. "For the most part," the creators wrote in their publication announcement in 2002, "researchers would prefer to view the genome at a higher level—the level of an exon, a gene, a chromosome band, or a biochemical pathway."[35] The genome browser was developed, between 2000 and 2002, from a small script that displayed a splicing diagram for the *C. elegans* worm into a powerful interface running a MySQL database over multiple Linux servers. Both Ensembl and the UCSC Genome Browser were developed in and for World Wide Web–based biology. The notion of a "browser" for navigating the genome borrows from the "web browsers" developed in the 1990s for navigating the Internet. A key consideration in the design of the UCSC browser was the need for speed in accessing data across the whole human genome. Unlike Ensembl, the browser was written in C, a low-level programming language that allows fast response times.

Programming in C, however, created a set of other problems that had to be overcome for the browser to be successful. As the designers put it, there is a "natural tension between how an object is represented in the database and in computer programs such as the scripts that make up the browser."[36] In other words, the programmers had to find a way of translating between "objects" in the relational database (rows in a table of data, for instance) and "structs" in the memory of a C program.

The solution was to create a special program, called autoSql, that could "talk to" both SQL (the database) and C (the programming language): providing autoSql with a structure for storing information automatically generates an appropriate database table for storing that information and appropriate C code for adding, retrieving, or deleting information from that table.[37] The genome browser must manage the relationships between five distinct representations of biological data: their representation as raw text stored in the database, their representation as an HTML/CGI request, their representation as an SQL database table, their representation as an object in C, and their representation as an image that is displayed on the screen to the user. When a user uses a web browser to navigate through the genome, the genome browser must deal with each of these in turn. The requested object appears as an HTML/CGI request, which is translated into a C script that generates a database query in the SQL language; the database returns a table of data, which is then parsed by more C code into text and images that can be displayed on a screen. In other words, the object in question undergoes a series of transformations: the genome morphs from a text of As, Gs, Ts, and Cs to a set of one-dimensional position coordinates in a database, to an array in a piece of code, to a picture on the screen.

The complexity of this system results from the fact that huge amounts of data need to be made accessible in fractions of a second. Storing the data, accessing the data, and displaying the data each have their own appropriate modes of representation. Doing bioinformatics means solving the practical problems of how to transform and manipulate these representations of biological objects into forms that can be used by biologists. Understood in another way, HTML, SQL, C, and so forth embody distinct logical domains in which objects must be structured according to particular rules; particular representations are ways of translating or mediating between those domains. Programs like autoSql generate boundary objects that can be passed back and forth.

But bioinformaticians solving such problems also face another set of even more important issues. Each of the representations of the genome must not only be interchangeable with the others, but also make sense of the genome as an object in the real world. Brian Cantwell Smith makes the point that the "choices one makes about the *content* of the various pieces of a representational system—how the representation represents its subject matter as being—are often affected by how one describes the computational structures themselves, the structures that carry or have content."[38] Decisions about how to represent an object entail decisions about what that object must be in the real world. Biological objects like genomes are not something that we can pick up and hold in our hands—we do not have an intuitive ontological grasp of what they are. Computational representations of a genome—for instance, what is part of the genome and what is an extraneous "label"—describe what a genome is.

The UCSC Genome Browser interface itself consists of a series of "tracks" displayed one below the other horizontally across the browser window (figure 6.4). The horizontal distance across the screen corresponds to the position in the genome, such that reading vertically down a column allows comparison of different features at the same genomic position. "The graphic display of the browser," the creators write, "is invaluable for getting a quick overview of a particular region in the genome and for visually correlating various types of features."[39] The tracks represent genes (as discovered by various methods), exons, gaps, expressed sequence tag alignments, cross-species homologies, single nucleotide polymorphisms, sequence tagged sites, and transposon repeats. Users can also add their own "custom tracks" to the image, which allows them to view, for example, how their own positioning of genes lines up with the standard annotation. Clicking on particular tracks leads to more detailed information: links to other databases or information

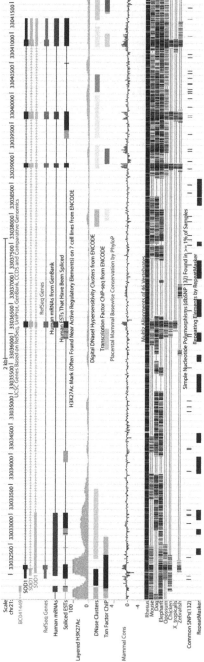

FIGURE 6.4 Screenshot of the UCSC Genome Browser. Note the similarity to AceDB in figure 6.2. Here, however, the genome is shown horizontally, with features such as mRNAs and ESTs displayed as parallel tracks. This format facilitates easy comparison with other aligned genomes (such as mouse, dog, elephant, etc. here). (http://genome.ucsc.edu/ [see Kent et al., "The Human Genome Browser"]. Reproduced by permission of University of California Santa Cruz Genome Bioinformatics.)

about the structure or function of proteins encoded by the clicked gene, for example. The controls at the top of the window allow the user to scroll along the genome or to zoom in on a particular position. One can also move around the genome by searching for particular genes or by typing in genomic coordinates. As such, the image is dynamically generated as the user navigates around the genome.

The UCSC Genome Browser is one of the most common ways in which biologists interact with the genome. It is an ideal tool with which to perform a quick check on a gene or region of interest (for instance, to see whether it is conserved, or to see whether it is aligned with any ESTs), or to systematically search for unidentified genes using multiple sources of information (for instance, by comparing computational predictions and conservation data). For example, a researcher might wish to know what genes lie in the first 500,000 base pairs of human chromosome 1. This can be done by searching the region "chr1:0–500000" for UCSC "knowngenes." The browser can then be used to examine whether particular genes are conserved in other vertebrates or to find the locations of SNPs, alignments with ESTs or GenBank mRNA, and a large amount of other information that associates particular genes with gene products, diseases, microarray data, gene ontology information, and metabolic pathways.[40] The presentation and organization of these data are visual—the correspondence between different gene predictions and between the organization and structure of genes in different organisms can be understood by visual comparison down the screen (see figure 6.4). The display has been described as follows:

> Gene structure is shown in these tracks with filled blocks representing exons; thick blocks are in the coding sequence (CDS) and thin blocks represent untranslated regions (UTRs). The lines connecting the blocks are introns. The direction of the arrowheads on the lines or the blocks show the strand on which the element resides.[41]

Elsewhere, colors are used to represent different sorts of codons as well as regions of high or low conservation.

The pictorial space generated from the database creates what Durbin and Thierry-Mieg call a "dense navigable network" that can be explored by pointing and clicking. Tools like the UCSC Genome Browser have generated a visual lexicon for thinking and talking about genomes. The genome has become a "map" on which the user can "navigate" around or "zoom" in. One of the advantages of AceDB was considered to be that it "permits biologists to describe and organize their data in a manner that closely resembles how they typically think about their information." Indeed, when using the genome browser, one has the impression of turning a microscope on the genome—as you zoom in far enough, you begin to see the tiny As, Gs, Ts, and Cs emerge in the tracks at the bottom of the window. This illusion of direct representation—the notion that we are really navigating around a one-dimensional space provided by the genome—no doubt contributes to the browser's usefulness. But the browser is also a tool for making the genome visible and legible: in the browser, the genome is totally visible and manipulable, it is a territory that is on display for quantification and analysis. Indeed, biologists often talk about using the browser for "data mining"—discovering regions of interest and extracting biological meaning and knowledge from them. The "illusory" quality of the browser is what gives it power—it allows the user to see hidden connections between tracks; it acts as a way of making biological knowledge by drawing objects into juxtaposition.[42]

Of course, the user is not actually "seeing" the genome: a genome is not even a string of As, Gs, Ts, and Cs, but rather a coiled-up set of macromolecules with some repetitive structural elements. Genes, exons, and binding sites are not labeled, not visible through any kind of microscope. Rather, the genome browser image is the result of an elaborate set of computations and manipulations of sequence data. Just as a political map shows imaginary boundaries that correspond to no visible feature on the ground, the browser allows biologists to inscribe categories and order on the genome in order to make sense of it. A track showing the positions of genes, for example, might reflect the output of a program like GenScan that uses a hidden Markov model to predict the positions of genes within a genome.[43] Tracks showing information about conservation between species are computed from painstaking attempts to align genomes with one another, processes that themselves require large amounts of computational time and power. Even the very positional information on which the genome sequence is based has been computationally reconstructed from millions of shotgun-sequenced pieces. The biological object is made visible through a series of re-representations and computations. The user, then, is not looking at the genome, but rather seeing the results of a series of analyses of biological information (both genetic and nongenetic). The picture the user sees is not a representation, but a computation.

The problem of translation between biology and computing is a problem of representation. It is in finding practical and computationally efficient solutions to this problem that the UCSC Genome Browser becomes a powerful tool for providing biological insight. The solution that the genome browser uses is primarily a pictorial one: its representation of the data as a dynamically generated set of images gives it the ability to show that which is of biological interest. The data themselves exist as a vast mass of hierarchical and cross-referenced tables within a computer. What the biologist wants is specific information about, for instance, the location or size or annotation of a gene. It is in and through the genome browser images that these two domains interact—that data can be extracted from the databases in ways that biologists can use. The genome browser is a visual solution to the problem of organizing and accessing large volumes of computational analysis in a biologically meaningful way. Moreover, solutions to the problems of how to represent objects within the computer and on the screen generate particular ways of seeing and interacting with biological objects; in an important sense, these computational representations reconstitute the objects themselves.

Many other examples could be given in which computational biologists use images to work with large sets of data or with computations. In genomics in particular, tools such as the UCSC Genome Browser are used to visualize genomes not just as sets of As, Gs, Ts, and Cs, but as one-dimensional maps showing the positions of chromosomes, genes, exons, or conservation. All these examples use machine-generated images to render the invisible visible. The representations are a way of seeing, organizing, and surveying large amounts of data. Their aim is both depiction of something "out there" in the biological world and something "in there," inside a computer. It is the representations themselves, in the images and in the databases, that stand between these two domains. These representations are valuable tools because they constitute objects in ways that allow them to be drawn into new and nonobvious relationships with one another.

Solving Problems with Pictures

None of these systems of representation are made up by biologists on the spot. As these examples show, they are embedded in the histories of software and hardware. But we also want to know how biologists actually go about solving problems by making pictures: What do they actually do with these images? This section describes the use of heat maps, a new tool for visualizing alternative splicing, and the work of a computational information designer. In these examples, reducing a

problem to an image provides not only a standardization or an aggregation of data, but also a way of seeing through the messiness (this was
often the way my informants described it) of biology to the hidden patterns in the data.

One of the most common images seen in contemporary biology papers is the heat map.[44] A typical heat map, like the one shown in figure 6.5, shows the correspondence between phenotypic or disease data
from patients and the levels of particular proteins reported in bioassays
of the same patients' cells. Here, the labels across the top correspond
to cell samples taken from patients with cancer, while the labels on the
left report the levels of particular proteins within their cells. Squares are
shaded lighter where the level of a protein is elevated above normal in
a particular patient, darker where the level is depressed, and black if
there is no measurable difference. Both the patients and the proteins are
ordered using a "clustering algorithm"—software that takes a multidimensional set of measurements and clusters them into a tree (shown
at the far top and far left of the heat map) according to how similar
they are to one another.[45] Heat maps summarize a vast amount of information; their convenience and popularity are due to the fact that
they provide a quick visual representation of multidimensional data.
Figure 6.5, for instance, presents the results of at least 538 (17×34)
distinct measurements.[46]

Heat maps are often used for DNA microarray data, where they provide the possibility of simultaneously displaying the effects of multiple
conditions (drugs, cell types, environmental conditions) on many genes
(all the genes spotted onto the array). Indeed, heat maps have become an
integral part of the presentation and understanding of microarray data.
In one of their early papers, the developers of microarrays discussed the
problems raised by the large data volumes produced by the arrays:

> Although various clustering methods can usefully organize ta-
> bles of gene expression measurements, the resulting ordered but
> still massive collection of numbers remains difficult to assimi-
> late. Therefore, we always combine clustering methods with a
> graphical representation of the primary data by representing
> each data point with a color that quantitatively and qualita-
> tively reflects the original experimental observations. The end
> product is a representation of complex gene expression data
> that, through statistical organization and graphical display, al-
> lows biologists to assimilate and explore the data in a natural
> intuitive manner.[47]

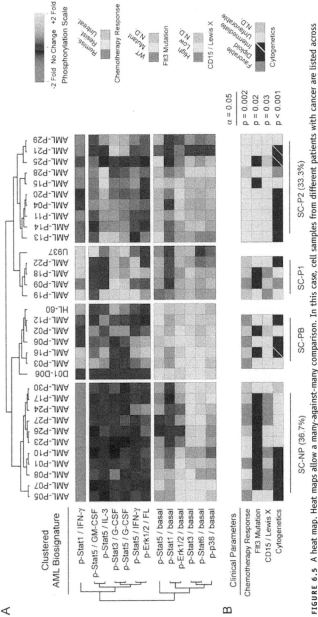

FIGURE 6.5 A heat map. Heat maps allow a many-against-many comparison. In this case, cell samples from different patients with cancer are listed across the top. Different environmental conditions (protein levels) are listed down the left side. The main map shows the protein phosphorylation response for each cell sample under each condition. The results are "clustered" (treelike structures on top and left) to show both cells that behave similarly and conditions that produce similar results in different cells. This clustering makes it straightforward to see large-scale patterns in the data. (Irish et al., "Single cell profiling." Reproduced by permission of Elsevier.)

Visualization provided a "natural" solution to the data management problem.

The history of heat maps is entrenched in the history of computers in biology. The heat map concept can trace its origins to attempts to apply computers to taxonomy.[48] Sokal and Sneath's methods of "numerical taxonomy" used computers to do the number crunching required to reduce large amounts of taxonomic data into a tree showing the "objective" relationship between organisms.[49] These techniques were developed into computer programs such as NT-SYS and SYSTAT that automatically produced high-resolution heat maps from a data set. Heat maps also have precursors in sociology: Wilkinson (the developer of SYSTAT) and Friendly trace the origin of heat maps to nineteenth-century tables measuring the geographic distribution of social characteristics (age, national origin, professions, social classes) across the Parisian *arrondissements*.[50] Just as such colored charts were designed to provide a visual overview of a city's population, heat maps are designed to provide a broad overview of a genome or other complex biological system. The clustering produces lumps of color that allow the biologist to immediately recognize patterns and similarities that would be difficult to glean from an unordered set of numbers.

Despite the seductive qualities of the image, biologists are aware of some of the limitations of such data reduction: "[A heat map] provides only first order insight into the data; complex patterns of non-linear relationship among only a few of the samples are unlikely to show up."[51] Perhaps more problematically, a heat map suggests ways to characterize particular biologies (e.g., cancerous cells versus noncancerous cells) based on just a few genes or markers; such clear-cut distinctions can be misleading, since they may be produced by random effects of arbitrarily selecting a few genes from among the twenty thousand or so in the human genome.

Nevertheless, the great abstraction and aggregation of a heat map makes it possible to do many things with biological data. Once again, the image is in no sense a direct representation of a biological object—it is based on a calculation, a manipulation. It allows the massive amounts of data processed by the computer to be communicated to the researcher in such a way that they can be made sense of and interpreted as biologically meaningful. But in so doing, it also imposes its own visual logic on biology—it is literally and figuratively a grid through which the underlying biology must be viewed. In particular, it illustrates simple, linear patterns, often clearly dividing samples into two, three, or four categories; other complicated and combinatorial relationships may be obscured.

During my fieldwork at the Burge lab, I observed firsthand the development of a new visualization tool. During the lab's work on the alternative splicing of mRNA, the group began to collect large amounts of new data on alternative splicing events.[52] In trying to come to terms with this new information, members of the lab realized that many of the basic questions about alternative splicing remained unanswered: What fraction of human genes can be alternatively spliced? How does alternative splicing relate to cell differentiation? Do different sorts of alternative splicing events play different biological roles? The new data consisted of over 400 million 32-base-pair mRNA sequences from ten different kinds of human tissues (skeletal muscle, brain, lung, heart, liver, etc.) and five cancer cell lines. This overwhelming volume of data came from the new Solexa-Illumina sequencing machines that had recently been made available to the laboratory.[53]

After mapping the sequences to the human genome, the group quickly discovered that almost 90% of human genes seem to undergo alternative splicing.[54] In order to answer the second and third questions posed above, however, the lab needed a way to summarize and organize the vast amounts of data in order to "see" how alternative splicing varied from tissue to tissue. A significant part of the effort to understand what role alternative splicing was playing was devoted to creating a visualization tool in order to capture the essence of the data. This was not just a process of finding a way to communicate information, nor was it a problem of how to present data at a conference or in a scientific journal article. Rather, it was a problem of organizing the data so as to see what was going on inside the cell.

The problem the group confronted was that their data had become multidimensional. To begin with, they needed something like a genome browser that laid out the linear order of the exons. However, in order to observe the different ways in which exons could be connected in different tissues, they also needed to find a way to visually differentiate those tissues and connections. Much discussion centered on the appropriate use of shapes, colors, and lines to achieve the desired representation. Some iterations of the tool brought spurious artifacts into view: "This representation tends to emphasize the weird events," Burge worried at one point. Early versions of the tool generated images that were difficult to understand because the data in the small exonic regions became lost in the larger, blank intronic regions and the plots displayed

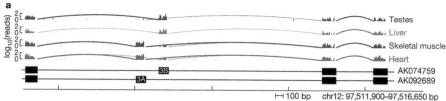

FIGURE 6.6 Visualization of alternative exon splicing. This image is similar to a genome browser image, displaying a region of sequence horizontally. In this gene, two possible splice patterns are observed, one including exon 3A but leaving out 3B, and the other leaving out 3A but including 3B. The image shows that the former splice pattern tends to occur in skeletal muscle and heart tissue (although not all the time) and the latter tends to occur in testis and liver tissue (although in testes, 3A is occasionally included). (Wang et al., "Alternative isoform regulation." Reproduced by permission of Nature Publishing Group.)

a large amount of meaningless irregularity. Lab members worked to tune the image so that only the salient features would be brought into focus.

Figure 6.6 shows an image produced by the visualization tool. The linear layout of the exons is displayed across the bottom, while four different tissues are displayed in different lines at the top (rendered in different colors in the original). Most significantly, arcs connect exons to one another to show the splicing pattern. In the case of this gene (AK074759), we can see that splicing in skeletal muscle and heart cells predominantly connects exon 2 to exon 3A to exon 4, while splicing in liver and testis cells predominantly makes the connection via exon 3B. Such an image provides clear evidence that, at least for this gene, the way in which exons are spliced could play a significant role in making a skeletal muscle cell different from a testis cell or a liver cell. It does so by summarizing and organizing vast amounts of data and computational work. The small graphs appearing over each exon show the number of sequences counted beginning at each nucleotide position. Since the scale is logarithmic, the whole image captures information gleaned from thousands of 32-base-pair Solexa-generated sequences. These sequences were then subjected to computational processing to map them to the human genome and to organize them into a "digital archive."

Coming to terms with this large, multidimensional data set meant producing a visual system. It was not the case that the lab members understood the role of alternative splicing in cell differentiation before doing the work to make it visible; knowledge about this feature of biology was produced through the development of the visualization tool.

It was in following this work that I discovered that some of the lab members were thinking seriously about the problems that visualization presented to biology. Indeed, Monica, a lab member working on the alternative splicing tool, hoped—upon gaining her PhD—to find a position working on visualization tools, rather than in a traditional biological laboratory. In an interview, she described her view of visualization tools in biology:

> I kinda believe that if people could explore their data in an unbiased fashion they might find interesting things. Often how we work as scientists is that you create some hypothesis, . . . and then we go look for statistical evidence of that. But if you could go and look at your data, you might actually see patterns . . . that you didn't have before. . . . What I am talking about is a tool that would facilitate people's discovery and facilitate people being actually able to explore and navigate their data.

For Monica, this was a new and exciting field, since the need to use such tools had arisen from the massive growth of biological data in the last few decades. Moreover, in her view, visualization was not about presenting or communicating data—making pretty pictures—but about creating images in a way that would help directly with research. Indeed, it had the potential to create an "unbiased" and hypothesis-free way of doing biology. Once again, this notion of "objectivity" is not based on using a computer to make a true-to-nature image, or "mechanical objectivity"; rather, it is about using visualization tools to create a kind of "free play" in which objects fall into juxtapositions that produce surprising but insightful results.[55] The images produced may be considered "unbiased" because they place objects in new structures and new relationships with one another. This sort of objectivity is based on the notion that the user-biologist can use the images to navigate around and connect up his or her data in an unconstrained way and hence be led toward perceiving true relationships.

Through Monica, I was introduced to a small group of people who circulated between biology and art, between the Broad Institute and the MIT Media Lab. They were all thinking about how to use images to represent biological data and solve biological problems. One of them, Benjamin Fry, completed his dissertation at the MIT Media Lab in 2004 and continued to work on problems of representing scientific data. His thesis, "Computational Information Design," called for the foundation of a new field that would bring together information visualization, data mining, and graphic design to address problems in the visualization of complex data.[56] In Fry's view, the need for images is a direct product of the growing volume of biological data:

The quantity of such data makes it extremely difficult to gain a "big picture" understanding of its meaning. The problem is further compounded by the continually changing nature of the data, the result of new information being added, or older information being continuously refined. The amount of data necessitates new software-based tools, and its complexity requires extra consideration be taken in its visual representation in order to highlight features in order of their importance, reveal patterns in the data, and simultaneously show features of the data that exist across multiple dimensions.[57]

Fry's images tend to incorporate interactive elements that draw the viewer/user into an exploration of the patterns in the underlying data. For instance, one of Fry's creations tracks the performance of major league baseball teams against their total salary expenditure.[58] By dragging a bar at the top of the image, the user can adjust the date at which the comparison is made, showing how teams' performances change over the course of the season.

In biology, Fry has developed a prototype of a handheld genome browser, a visual representation of the BLAST algorithm (called "Genome Valence"), software for generating a three-dimensional axonometric display of genes on a chromosome, a tool for understanding haplotype data, a three-dimensional genome browser (called "Strippy"), and a redesigned two-dimensional genome browser.[59] The main example elaborated in his thesis is a sophisticated tool for viewing genetic variation between individuals, which combines haplotype data (SNPs) with linkage disequilibrium data. The user is able to view the data in multiple ways; clicking on different parts of the image exposes new information or more details. "The user can easily transition between each type of view. . . . The software is built to morph between representations, providing a tight coupling between the qualitative—useful for an initial impression and getting a "feel" for that data—with the quantitative— necessary for determining specific frequencies of haplotypes of interest for specific study."[60]

In describing his concept of computational information design, Fry argues that the role of the designer is not to create complex images that summarize all the data, but rather to find out which data are necessary to answer a particular question and to make sure that they stand out in the image. He uses the example of the London Underground map, designed in the 1930s, to show how simplification to horizontal, vertical, and diagonal lines, showing only the transfer points between lines, allowed commuters to quickly figure out how to get from A to B. "Often less detail will actually convey more information, because inclusion of overly-specific details cause the viewer to disregard the image because of its complexity."[61] Indeed, the recurring principles in Fry's work are simplicity and clarity: "How can this be shown most clearly, most cleanly?"[62] Fry has also built a software platform, a sort of visual programming language, for scientists to use in generating visualizations.[63] This program, called "Processing" (developed by Fry and Casey Reas), is a way of reorienting programming from text to images:

> Graphical user interfaces became mainstream nearly twenty years ago, but programming fundamentals are still primarily taught through the command line interface. Classes proceed from outputting text to the screen, to GUI, to computer graphics (if at all). It is possible to teach programming in a way that moves graphics and concepts of interaction closer to the surface. The "Hello World" program can be replaced with drawing a line, thus shifting the focus of computing from ASCII to images and engaging people with visual and spatial inclinations.[64]

In shifting programming from ASCII to images, Fry is also attempting to transform biological investigation from AGTC to images. Fry understands his representations as means of simplifying the messy terrain of biological data into maps that can be easily used to comprehend and manipulate biology. These images are not intended to be printed out or displayed in academic journals—they are thoroughly embedded in the computational infrastructure that generates them and, to be useful, must be continually engaged by the user in a process of interaction and discovery.

In Fry's work, we see how seriously and deeply intertwined problems of visualization and biology have become. In particular, we see how the image becomes a way of doing biology computationally—Processing is a tool both for doing biology (making images for data analysis) and for doing programming. It is these innovations in visualization that are driving some of the most novel work in bioinformatics. Fry's (and others') tools allow more and more data to be digested, ordered, reordered, and analyzed. It is through them that biologists are able to "see" what is going on amid the masses of genomic and other data, using color, shade, shape, and size to expose hidden patterns.

Conclusions

One of the key differences that computers have made to science is that they have enabled a new way of seeing.[65] Like the microscope or the telescope or almost any other scientific instrument, the computer can be understood as a visual aid. The visual realm—pictures of proteins and sequences—provides a way of mediating between the material and the virtual.[66] In the most obvious sense, computers are unable to work directly with biological materials—they can only manipulate numerical or digital distillations of macromolecules or cells. Conversely, numbers or digital codes do not have inherent meaning—they must be reinterpreted into the language of biology. Visualizations mediate between the digital and the bio-logic and thus play a crucial enabling role in bioinformatics.

The ability of images to have meanings (even if such meanings are not identical) for both biologists and computer scientists is no doubt part of the explanation for their ubiquity. Biological ways of knowing the world and digital ways of knowing the world are brought into dialogue, not through common training or common language, but through the image. Images constitute new ways of making biological knowledge that draw biological entities into new relationships with one another—out of sequence—to reveal hidden patterns. Sequences are constituted not only as ordered, linear text, but also as reordered, multidimensional grids and maps. Image making is not only about motion and fluidity, but also about moving things around in order to identify durable and robust relationships between objects; images help to stabilize knowledge claims.

The creation of images using computers also constitutes a new mode of biological practice. One of the features of bioinformatics is its ability to move biology away from the laboratory bench into a statistical, mathematical, or even theoretical space that is dominated not by cells and chromosomes, but by chips and codes. An "experiment" can be run without even leaving one's desk, or without ever having to touch a cell or a pipette. One of the ways in which such experiments take place is by making and viewing images. Much of the work that has addressed the role of computers in biology has described the computers themselves, their software, and their histories.[67] Less has been said about what biologists actually *do* when they are doing biology in front of a computer screen. This chapter has provided some examples: they are browsing around genomes, they are building and examining heat maps, they are scrutinizing visual representations of sequence alignments, they are using images to reorder, interrogate, and find patterns in their data.

Images also have power as metaphors. Just as texts and codes bring with them a powerful set of metaphors for biology, images too engender particular ways of looking at living substance. Because images often aim to simplify the object they are reconstructing, they perform a kind of visual synecdoche: the distances between specific atoms in a protein molecule substitute for the complicated three-dimensional structure of the whole. Or, in the kind of metaphor we have seen here, the genome or protein becomes a map, a space to be navigated around in one or two dimensions. Privileging spatial relationships between parts of proteins or genome sequences is just one way of thinking about these biological objects. Relationships that are dynamic, temporal, biochemical, or network-oriented are not compatible with seeing the protein or genome as a static space that can be navigated around. The map metaphor suggests that biology works only or largely by "navigations"—by finding "pathways" or "road maps" through the landscapes of genes and proteins. These maps, trees, charts, and matrices also provide control—they are ways of rendering the territory of biology as an orderly and disciplined space that can be manipulated and intervened in.[68]

Visualizations are not intended to be "realistic" depictions of biological objects, but rather to capture particular features of those objects in order to be able to translate successfully between the computational and the biological realms. Genomes—as bioinformaticians (and an increasing number of biologists) know them—are necessarily mediated through these computational representations. It is the structures of the data that come to constitute the objects themselves. As Brian Cantwell Smith makes clear, making computational objects is never just an abstract enterprise in computer programming; rather, it is "computation in the wild"—that is, it is making objects in the world.[69] Genomes are not totally informatic constructions—their form still refers back to the molecules from which they derive. But the representational and visual constructs of bioinformatics—including the database spaces examined in chapter 5—impose structures and constraints on biological objects; these structures allow them to interact with certain objects and not others, allow them to take certain shapes and not others, allow them to be manipulated in certain ways and not others.

Conclusion: The End of Bioinformatics

We have followed the data: out of physics, through the spaces of laboratories, through networks and databases, and into images and published scientific papers. This approach has allowed us to see how biological work and biological knowledge is shaped and textured by digital infrastructures. Rather than just being speeded up or liquefied or virtualized, we have seen how data, along with the structures they inhabit, tie biology to the everyday practices of biologists and the computational tools they use, albeit in new ways. Biological knowledge now depends on virtual spaces, production lines, databases, and visualization software.

In the end, what difference does all this make to our understanding of life? In the first place, it has diverted the attention of biologists toward sequence. This is not as simple a claim as it may at first seem. Computers are data processing machines, and sequences have come to the fore by virtue of their status as data. The notion of sequence has become inseparable from its role as data; it must move and interact and be stored as data. The ubiquity of data has carved out a new significance for statistics in biology: computers imported statistical approaches from physics that have allowed sequences to be treated statistically and to be manipulated, rearranged, and related to one another in ways not possible without computers. Sequences have been reconceived as objects "out of sequence," or "out of order," making way for techniques of randomness and stochasticity to be applied to analyzing genes and genomes.

Understanding the connections between sequences, statistics, and computers can help us make sense of recent pronouncements about biology, such as this one from Craig Venter in 2012:

> We're at the point where we don't need one genome or just a few genomes to interpret your genome. We need tens of thousands of genomes as a starting point, coupled with everything we can know about their physiology. It's only when we do that giant computer search, putting all that DNA together, that we will be able to make sense in a meaningful statistical manner of what your DNA is telling you. We're just at the start of trying to do that.[1]

With enough sequence data and powerful enough computers, many biologists believe, it will be possible to answer almost any question in biology, including and especially big questions about whole organisms, bodies, and diseases. What "your DNA is telling you" depends on its data-based relationships to thousands or millions of other pieces of sequence. So the turn toward sequence is not just reductionism writ large, or a scaling up and speeding up. It also signifies a faith in particular tools (computers), particular approaches (statistics), particular objects (data), and particular types of questions (large scale) to move biological knowledge forward.

More specifically, "what your DNA is telling you" doesn't just depend on your genome or the thousands of other genomes out there. Rather, it depends on how they can be related to one another; these vast amounts of data can be made sense of only by using computers to search for specific patterns and relationships. These relationships, in turn, depend on the structures into which the data are fitted: the ontologies, the databases, the file structures, and the visual representations into which they are poured. It is these structures—written into hardware and software—that order and reorder sequence data such that biologists can find meaning in them. It is the rigidity of these forms that this book has sought to describe and understand. Databases, ontologies, and visual representations tie informatic genomes to the specific practices of computers, computational biology, and bioinformatics.

Our understanding of our own genomes also increasingly depends on vast amounts of *other kinds* of data that are being rapidly accumulated. This book has focused on genomes themselves, and especially on human genomes. But human genomes are increasingly understood by comparing them with the genomes of the hundreds of other fully sequenced organisms—and not just organisms that are perceived to be similar to humans (such as apes and monkeys). Rapid sequencing and data analysis have made it possible to sequence the thousands of organisms that inhabit the human gut, the oceans, the soil, and the rest of the world around us. In 2012, the Human Microbiome Project reported that each individual is home to about 100 trillion microscopic cells, including up to ten thousand different types of bacteria.[2] The Earth Microbiome Project estimates that we have sequenced only about one-hundredth of the DNA found in a liter of seawater or a gram of soil.[3]

Many such studies are based on "metagenomic" approaches, in which sequencing is performed on samples taken from the environment without worrying about the details of which sequence fragments belong to which organisms. By showing our dependence on a massive variety of microorganisms living on us and around us, metagenomics is providing a new perspective on biology. These techniques would be not only impractical, but also incomprehensible, without the databases, software, and hardware described here. It is the possibility of using software to pick out genes in fragments of sequence, and the possibility of storing and sharing this information in worldwide databases, that makes this new field make sense.

Further, there are vast amounts of *non-sequence* data being gathered about human biology as well. These include data on proteins and their interactions (proteomics, interactomics), metabolites (metabolomics), and epigenetic modifications to genomes (epigenomics). High-throughput techniques such as yeast two-hybrid screening, ChIP-chip (chromatin immune-precipitation on a microarray chip), ChIP-seq (chromatin immunoprecipitation and sequencing), and flow cytometry produce large amounts of nongenomic data. Understanding our genomes depends on understanding these other data as well, and on understanding how they relate to, modify, and interact with sequence data. However, part of the role that bioinformatics tools play is to structure these other data *around* genomes. We have seen how databases like Ensembl and visualization tools like the UCSC Genome Browser use the genome sequence as a scaffold on which to hang other forms of data. Sequence remains at the center, while proteins and epigenetic modifications are linked to it; the genome provides the principle on which other data are organized. This hierarchy—with sequence at the top—is *built into* the databases, software, and hardware of biology.

Where is sequence likely to take biology in the near future? This conclusion is an apt place to make some suggestions. Sequence is not going away: next-generation sequencing machines are making more and more sequence and more and more data an increasingly taken-for-granted part of biology. The ways in which these increasingly massive amounts of data are managed are likely to become ever more entangled with the management of data in other domains, especially with web-based technology. Bioinformatics will become just one of many data management problems. These changes will have consequences not only for biological work, but also—as the results of bioinformatics are deployed in medicine—for our understanding of our bodies. Computational approaches to biology may become so ubiquitous that "bioinformatics"—as distinct from other kinds of biology—will disappear as a meaningful term of reference.

Next Generations

Biological materiality is becoming ever more interchangeable with data. Since 2005, so-called next-generation sequencing machines have given biologists the ability to sequence ever faster and ever cheaper. In the medium term, as these machines are sold to hospitals and companies selling personalized genomic tests, this phenomenon is likely to have a profound effect on medical care.[4] However, these machines are also likely to find more use in basic biology, and here too, they are likely to have profound effects on the production of biological knowledge. This study of bioinformatics—in many ways a study of the speeding up of biology—should prepare us to understand what the consequences of next-generation sequencing may be.

Next-generation technologies (also called next-gen) were inspired not only by the HGP—which demonstrated the commercial potential of sequencing machines—but also by data-driven biology. If bioinformatics, as we have described it here, is about high volume and high throughput of data (especially sequence data), then next-gen is its technological complement. Both the technology and the science itself can be said to be driven by data. Each provides a further justification for the other: we need more and faster sequencing technology so we can amass more data, and we need the statistical techniques of data-driven science in order to digest those data into useful knowledge. So bioinformatics and next-generation technologies go hand in hand and stand in a relationship of mutual causation.

Moreover, next-generation sequencing seems to lie at a crucial intersection of computing and biology. Touted—over and over again—as the embodiment of "Moore's law for biology," next-gen is literally and figuratively connecting the history of semiconductor computer chips to biological progress.[5] Graphs and charts depicting the logarithmic increases in sequencing speed and decreases in sequencing cost suggest that biological progress will proceed as inexorably as the increases in computing power since the 1960s predicted by Moore's law. Jonathan Rothberg's Ion Torrent Personal Genome Machine makes this connection directly: this recent piece of next-generation technology uses tiny pH meters embedded in wells on the surface of a silicon computer chip to detect the incorporation of a nucleotide. This allows the chip—which can be manufactured in the same way as any other microchip found in a desktop computer—to act as the basis for a sequencing machine.[6] Because it is built on standard computer chip technology, the Ion Torrent machine hitches biological progress directly to progress in the manufacturing of semiconductors (getting more transistors on a chip will allow more pH meters on a chip, meaning that sequencing can be done faster). The sequencing pipeline that we encountered in chapter 4 is rapidly being compressed—in time, space, cost, and complexity. The material and the digital seem to be moving closer and closer together.

This compression may have several consequences. First, the digital traces of sequences will come to seem more and more substitutable for biological samples themselves. If samples or specimens or biological materials can be rapidly and cheaply digitized, then it becomes even easier to replace the material with the virtual. We may pay even less attention to that which is left out of these digital traces, that which is stripped away in reducing samples to sequence. The constraints and structures imposed by computer and network tools will become even less obvious. Bioinformatics itself may be erased: the biological will become the informatic.

But this may be too simplistic (and perhaps too pessimistic) a view. Next-generation sequencing also opens up possibilities for reversing some of the trends we have encountered in this book. The most obvious possibility is that the organization of biological work may once again radically shift. As we have seen, the HGP inaugurated massive changes in the labor and economy of biology—these changes were necessary because the project required large amounts of money, large labs, interdisciplinary work, and centralized coordination over many years. But next-generation sequencing could allow genomics to be performed on a bench top, with a couple of operators, and over a few days or weeks. Whole genomes could be rapidly sequenced by a single lab, at a modest cost, and within the time frame of a small experiment. In other words, next-gen may have the effect of recentering biological work on the individual investigator and the individual lab. The Personal Genome Machine may make biology "personal" once again.

Anne Wojcicki of 23andMe has spoken about how rapid and large-scale sequencing projects might overhaul the biomedical research process. Rather than writing research grants and waiting for funding from the NIH, 23andMe aims to become a "research engine" in its own right. The amount of genomic data already accumulated from its customers allows researchers at 23andMe enough statistical power to pose and answer significant biomedical questions. By surveying its customers online about their habits, diseases, or family histories, 23andMe can rapidly generate statistical associations between phenotypes and genetic loci.[7] This kind of crowd-sourced genome-wide association study poses a range of ethical dilemmas,[8] but it once again suggests the potential of cheap and large-scale sequencing to transform research practices: large-scale sequencing centers requiring massive funding and centralized planning might give way to cheaper, more widely distributed forms of research. The who, where, and how of research is shifting rapidly.

This observation also reminds us that next-generation sequencing is—for the most part—strongly driven by commercial considerations (the exception is the work of George Church at Harvard Medical School). It is firmly a part of the highly competitive biotech world. Therefore, it also generates the possibility for further entanglement of commercial enterprise with basic science—commercial considerations could become central to the biological enterprise, structuring the types of work and the knowledge made from it. Next generation sequencing could contribute to the creation of the new kinds of value and new kinds of accounting that I have tracked here.

But next-gen may also entail shifts in how biologists understand organisms. One possibility is that it may allow biologists to understand the genome as a more dynamic object. At least at present, next-generation machines are used largely to generate short reads of DNA and RNA, providing snapshots of all the species of sequence inside a cell at a particular instant. Rather than simply allowing the sequencing of more and more complete genomes, next-gen is opening up possibilities for seeing cellular processes in action. Combined with the bioinformatic tools that have been described here, this shift opens up more possibilities for drawing genomic, transcriptomic, and epigenomic elements into new relationships with one another, studying them not as static elements, but as dynamically interacting parts. In other words, next-gen may generate a biology that remains focused on sequence, but on sequence "out of sequence"—that is, one that pays less attention to the static, linear genome and more to the patterns of multiple, dynamic, rearrangeable, lively interplay of its parts.

Next-generation sequencing may be "bioinformatic" not just because it produces huge amounts of data, and not just because it brings material and virtual biology closer together, but also because it helps to generate just the kinds of new relationships and patterns that we have seen in bioinformatic software, databases, and visualizations. In particular, next-generation sequencing reminds us that biology is increasingly driven by the exigencies of data. It is not just that biology is becoming information, but rather that biology is coming to depend more and more on the technologies and information structures that store and move data.

Biology and the Web

At the beginning of my fieldwork, I was consistently surprised by the ways in which biologists narrated the history of bioinformatics alongside histories of the Internet and information technology. An example appears in figure C.1. The first DNA sequence appears alongside the invention of email, and the development of sequence-matching algorithms (for instance, BLAST) alongside "Sendmail" and HTTP. Why would biologists, I wondered, think of their past in this way? Further investigation revealed not only that, from the origins of the ARPANET to the creation of cgi.pm, technical developments in bioinformatics influenced the developing World Wide Web, but also that biology was in some small part responsible for the Internet's explosive growth in the decade after about 1993. As access to large amounts of biological data (particularly sequence data) became a more and more necessary part of biological practice, biologists demanded that academic campuses have bigger, better, and faster Internet connections. In an interview in 2001, Ewan Birney, one of the pioneers of the EMBL database, when asked about the influence of the Internet on genomics, replied, "It's a two-way street—a lot of the web development was actually fostered by bioinformatics. . . . The needs of molecular biologists was probably one of the reasons why many campuses have upgraded their Internet connectivity over the last decade."[9]

It is biology that continues to push the limits of the Internet's capability for data transfer. For instance, GenBank and the EMBL database share large amounts of data in order to make sure their databases are

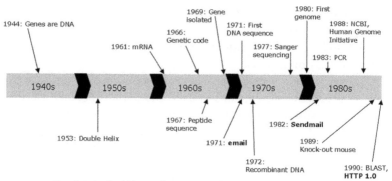

FIGURE 7.1 Time line showing biology and Internet milestones side by side. Similar time lines can be found in many introductory bioinformatics and computational biology courses or in online accounts of the history of bioinformatics. (Compiled by author from fieldwork notes.)

concurrent with one another. This sharing requires the daily exchange of raw DNA data dumps (swap trace files of terabyte size) across the Atlantic. Likewise, large sequencing centers, which are contractually required to submit their sequences to the appropriate databases on a daily basis, must often transmit massive amounts of data over the Internet every night. One computational biologist responsible for such uploads at the Broad Institute told me that some data dumps had been interpreted by Internet administrators as malicious BitTorrent attacks and subsequently shut down.

One specific example of the importance of biology to the Internet is the story of cgi.pm. In the mid-1990s, at the Whitehead Institute in Cambridge, Massachusetts, Lincoln Stein was working to publish the laboratory's physical maps of the genome. Stein needed a quick and cheap way to retrieve data from a database and display them on a web page. A few years earlier, Internet developers had created the Common Gateway Interface (CGI), which provided a language in which host machines could make requests to the server (to open a file on the server, for example). In 1995, Stein developed a module in the scripting language Perl called cgi.pm, which permitted information to be passed backward and forward quickly between host and server using CGI; cgi.pm takes user input, executes the appropriate operations on the server based on that input, and creates HTML output. The module became the basis for creating "dynamic" web pages, transforming the web from a fixed collection of pages into a space in which pages could generate their own content depending on user input.[10] The module became one of the most ubiquitous and widely used scripts on the Internet. It was cgi.pm that allowed the Internet to be transformed from a static, server-oriented entity into a flexible, user-centered tool. The notion that the user could send information over the Internet (for instance, a name, address, and credit card number) that could then be stored on the server or used to generate customized HTML pages was crucial for the development of e-commerce.[11]

Indeed, the Perl language more generally has had a profound effect on both the web and biology. In the early days of the web, Perl played the dominant role in CGI programming, which allowed data to be passed from a client machine to a program running on a server. Hassan Schroeder, Sun Microsystems' first webmaster, remarked, "Perl is the duct tape of the Internet."[12] Its ability to "glue together" programs written in a variety of other languages made Perl an ideal tool for the web developer, webmaster, or systems administrator. Similarly, it was Perl's ability to act as an adhesive between different sorts of programs that allowed it to "save the Human Genome Project." As Stein relates, the various labs working on sequencing the human genome each had their own independent methods of processing their data, so the resulting sequence data existed in a variety of incompatible formats.

> Perl was used for the glue to make these pieces of software fit together. Between each pair of interacting modules were one or more Perl scripts responsible for massaging the output of one module into the expected input for another. . . . Perl has been the solution of choice for genome centers whenever they need to exchange data, or to retrofit one center's software module to work with another center's system.[13]

Indeed, Perl became such a popular tool for biological work that developers at major genome centers created an open source community (called the BioPerl Project) for the purpose of developing and sharing Perl modules useful to biology. The ubiquity and importance of the Perl language in the development of the Internet and of the HGP shows that both projects faced similar obstacles that could be overcome with similar tools. Specifically, Perl provided ways in which small and powerful software components could be seamlessly linked together in order to facilitate the smooth transfer of data.

Bioinformatics and the Internet continue to face similar problems and continue to develop in parallel. In particular, the futures envisioned for the Internet remain deeply intertwined with efforts to understand biology, and especially the human genome. Writing about the "past, present, and future" of the web in 1996, Tim Berners-Lee looked forward to an Internet where not only humans, but also machines (that is, the computers used to access the web), would be able to understand its content.[14] If machines could understand the Internet, its power to enable the sharing of knowledge would be greatly enhanced. The problem, as Berners-Lee saw it, was that people "go to search engines and they ask a question and the search engine gives these stupid answers. It has read a large proportion of the pages on the entire Web . . . but it doesn't understand any of them. It tries to answer the question on that basis. Obviously you get pretty unpredictable results."[15] The solution he proposed was a "Semantic Web," in which the Internet would be written in a common, extensible, machine-understandable language. Berners-Lee imagined an Internet in which not only documents, but also data, would exist in a standardized format that would allow links (an HTML equivalent for data). "The Semantic Web," Berners-Lee believed, "will enable better data integration by allowing everyone who puts individual items of data on the Web to link them with other pieces of data using standard formats." Significantly, he took his first example from the life sciences:

> To appreciate the need for better data integration, compare the enormous volume of experimental data produced in commercial and academic drug discovery laboratories around the world, as against the stagnant pace of drug discovery. While market and regulatory factors play a role here, life science researchers are coming to the conclusion that in many cases no single lab, no single library, no single genomic data repository contains the information necessary to discover new drugs. Rather, the information necessary to understand the complex interactions between diseases, biological processes in the human body, and the vast array of chemical agents is spread out across the world in a myriad of databases, spreadsheets, and documents.[16]

In particular, Berners-Lee advocated RDF (Resource Description Framework) and Web Ontology Language (OWL) for standardizing data and connecting sources to one another. RDF is a set of specifications, or a "metadata model," that is used to label web pages with metadata. In other words, it aims to provide a common, cross-platform way of labeling web pages with information about their content. Because RDF format and terms are tightly controlled, these labels can be "understood" by a computer and can be processed and passed around between applications without meaning being lost.[17] For instance, a Wikipedia article about Lincoln Stein might have the following RDF:

```
<rdf:RDF>
<rdf:Description rdf:about="http://en.wikipedia.org/wiki/Lincoln_Stein">
<dc:title>Lincoln Stein</dc:title>
<dc:publisher>Wikipedia</dc:publisher>
</rdf:Description>
</rdf:RDF>
```

The tag structure, which works like HTML code, tells the computer that this web page is an article about Lincoln Stein published by Wikipedia.

The OWL language goes beyond RDF by including not only a vocabulary, but also a formal semantics that allows terms to be unambiguously related to one another. These relationships provide the basis for the possibility that machines can perform "useful reasoning tasks" on web documents.[18] OWL's developers explained its necessity as follows:

> The World Wide Web as it is currently constituted resembles a poorly mapped geography. Our insight into the documents and capabilities available are based on keyword searches, abetted by clever use of document connectivity and usage patterns. The sheer mass of this data is unmanageable without powerful tool support. In order to map this terrain more precisely, computational agents require machine-readable descriptions of the content and capabilities of Web accessible resources.[19]

Just imagine, they continued, asking a search engine, "Tell me what wines I should buy to serve with each course of the following menu. And, by the way, I don't like Sauternes." Or designing software to book coherent travel plans. The relationships in OWL would allow software to "understand" that a wine identified as a "merlot" must also be a red wine. The power of the Internet is unrealized, Berners-Lee and others claimed, because the massive amounts of data are not connected or organized in any systematic way. RDF and OWL are attempts to solve this problem by imposing a controlled vocabulary and semantics on the description of online objects (data, web pages).[20] They are "ontologies" because they aspire to a complete description of all things that exist on the web and the relationships between them.

As we saw in chapter 4, biology has paid significant attention to its "ontological" problems, and a wide range of biomedical ontologies already exist. These include ontologies for naming proteins, for the anatomy of ticks, for microarray experiment conditions, for clinical symptoms, for human diseases, for fly taxonomy, for metabolic pathways, and for systems biology (just to name a few).[21] Most of these ontologies are also OWLs. Indeed, many of the most sophisticated attempts to move toward the Semantic Web involve biological applications. For instance, a team at the Cincinnati Children's Hospital Medical Center employed a Semantic Web consultant to help them discover the underlying causes of cardiovascular diseases. Gathering data from a variety of biological databases, the team used open source Semantic Web software (Jena and Protégé) to integrate them into a common format. "The researchers then prioritized the hundreds of genes that might be involved with cardiac function by applying a ranking algorithm somewhat similar to the one Google uses to rank Web pages of search results."[22] Pharmaceutical companies, including Eli Lily and Pfizer, have also been involved in Semantic Web development efforts directed at drug discovery. The data richness of biology has made it the primary field for testing new Semantic Web technologies.[23]

Bio-ontologies, however, not only entailed a new set of language rules, but also fostered new ways of thinking about how to represent and interrelate information on the web. In the late 1990s, it became clear that understanding the human genome would mean not only building long lists of sequences and genes, but also understanding how they interacted with one another: biology shifted from static sequences and structures to dynamics. Deciphering gene networks and metabolic pathways became crucial. In May 2000, Eric Neumann, Aviv Regev, Joanne Luciano, and Vincent Schachter founded the BioPathways Consortium with the aim of promoting research on informatic technologies to capture, organize, and use pathway-related information. In the 1990s, Neumann had worked on biological problems using semantics, XML, and HTML tools. When Neumann encountered the problem of representing molecular pathways, he was reminded of Berners-Lee's Semantic Web ideas:

> [They] especially made sense in the context of trying to put together molecular pathway models, which are really semantically-defined graphs. In 2000 we took a stab at representing these using the N3 short-hand, and loading them into his [RDF]-tool. From this point on, I was a Semantic Web convert.[24]

Looking for a data-graph storage system for his company's informatics needs, Neumann began a collaboration with Oracle to test its new graph storage technology (called NDM). The success of this collaboration prompted Neumann to suggest that Oracle contact Berners-Lee, who was also looking for a data-graph system for the Semantic Web. This connection between Oracle and Berners-Lee led to the development of Oracle's RDF product, which has had a significant effect on the spread of the Semantic Web. As Neumann writes, "The connection made in trying to solve a life sciences data problem may have helped to jump start the recent Semantic Web movement."[25]

This intersection between biological work and web work occurred because the problems were made congruent. Biologists needed to find ways to draw heterogeneous data elements together into network representations. Some way of representing a pathway between proteins in different databases, for instance, had to be found. Neumann, Berners-Lee, and others came to conceive the problem of the Semantic Web in the same way: as a problem of representing networks of relationships between data on different websites.

As the new field of systems biology took off in the early 2000s, the analogy between biological and electronic networks became even more explicit. Systems biologists became concerned with gleaning biological insights not from the details of specific biological interactions, but from the large-scale properties of whole networks. Systems biologists demonstrated, for example, that protein interaction networks, like the Internet, were "small-world, scale-free" networks.[26] In 2004, Albert-László Barabási (who had discovered the scale-free properties of the web) and Zoltán Oltvai argued that the similarity between technological and biological networks could be explained by their common patterns of growth:

> Nodes prefer to connect to nodes that already have many links, a process that is known as preferential attachment. For example, on the World Wide Web we are more familiar with the highly connected web pages, and therefore are more likely to link to them. . . . [Similarly,] highly connected proteins have a natural advantage: it is not that they are more (or less) likely to be duplicated, but they are more likely to have a link to a duplicated protein than their weakly connected cousins, and therefore they are more likely to gain new links.[27]

By developing the analogy between the Internet and biological networks, these authors could gain insight into the way protein networks were built up through gene duplication events. Although there has now been some doubt cast on the idea that biological networks are truly scale-free,[28] the similarity between computer-electronic and biological networks suggests another reason why the same sets of practices and problems should apply to both.

During 2007 and 2008, the advent of various "personal genome projects" further imbricated biology and the web. Companies such as 23andMe and Navigenics began to operate genotyping services that made vast amounts of personal genomic and medical data available over the Internet. The web makes it possible not only to display these data in ways that nonspecialists can understand, but also to integrate them with social networking facilities (such as 23andWe, for those who wish to connect with other people with similar genetic traits), dietary advice, and exercise regimens (to offset risks of genetic predispositions), all downloadable to your iPhone or Blackberry.

In September 2007, I interviewed a software entrepreneur who had been working in biology for the last decade—he had worked with Eric Lander at the Whitehead Institute to produce some of the first detailed computational analysis of the human genome. His latest start-up was a professional networking website for scientists—for him, this was a natural step not only because it promoted the sharing of biological information, but also because it harnessed his own skills and experience in managing complex networks of data. For such entrepreneurs, and for companies like 23andMe, biological data, browsing the web, and social networking have become part of a single experience of sharing and using information.

In May 2007, Google invested $3.7 million in 23andMe. Notwithstanding the fact that Anne Wojcicki, cofounder, major shareholder, and member of the board of directors of 23andMe, was then engaged (later married) to Sergey Brin (one of the search engine's founders and president for technology), Google wanted to make sure that it had a stake in this new information revolution. As one Silicon Valley blogger put it, "If Google wants to really organize the world's information, it needs to consider DNA, the most personal of data."[29] For about $1,000, it may soon be possible to "Google your genome," searching the mass of available data just as one would search for a web page. The alliance between Google and 23andMe represents a merging of the biotech and web industries: 23andMe is based in Silicon Valley, just down the road from Google, and "feels very much like the quintessential startup."[30]

Not only Google, but Microsoft and a range of other software and web-centered companies as well, have now begun to take an interest in biomedicine. For example, Google set up Google Health (launched in

May 2008 and shut down January 2013) for managing personal health records, and Microsoft has HealthVault (launched in October 2007).[31] Google also offers "universal search in life sciences organizations"— "All the information you need to be productive at work should be available through a single search box—including all the content within your enterprise firewall, on your desktop, and on the World Wide Web." Google's search technology, the company promises, will be embedded seamlessly in the research and development effort, saving time and money.

From one point of view, Google's foray into the life sciences is unsurprising: it has designed tools to search and manage large amounts of information, and the proliferation of biological data offers it a natural market. On the other hand, seeing these two distinct kinds of objects (websites and biological data) as essentially similar is a nontrivial step: Why should information from genomes be similar enough in organization to the information from the World Wide Web that the same tools can apply to both? Why should "Googling your genome" be a thinkable practice? I am suggesting here that it is because biology and the web share both a history and a set of problems—because they are entangled with one another to such an extent that understanding the future of biology requires understanding where the Internet is headed as well. As biological knowledge is increasingly produced in and through the web, biological objects are increasingly online objects—the web representation of a genome or a cell becomes the way it is understood. Googling your genome becomes an obvious practice because one's genome is—first and foremost—data. Google's solutions to problems of data management are also solutions to biological and biomedical problems.

Biology 3.0

In 2006, the *New York Times* journalist John Markoff coined the term "Web 3.0" to refer to a collection of "third generation" web technologies, such as the Semantic Web, natural language searching, and machine learning, that he predicted would take over the Internet between 2010 and 2020. Although the terms are fuzzy and contested, Web 0.0 is usually understood as the predigital network of direct human-to-human interaction, unmediated by technology, and Web 1.0 as the first iteration of large-scale data sharing, mostly using HTML, in the 1990s. The first decade of the twenty-first century became dominated by Web 2.0, the use of the Internet for social networking, wikis, and media sharing. Web 3.0, in contrast, promises an "intelligent Web," in which data will be immediately accessible without searching and software will be able to extract meaning and order from the web and how people use it.[32] The idea of Web 3.0 is based on a set of technologies that are just beginning to emerge, especially the use of the Semantic Web to create a seamless space of interaction in which the web itself becomes able to "read" and "understand" its own content. Navigating around in Web 3.0 promises to be a process of both human and computer cognition—a further erasure of the boundary between minds and machines.

Others see Web 3.0 as allowing increased use of three-dimensional spaces, an apotheosis of Second Life, which "opens up new ways to connect and collaborate."[33] Yet more provocatively, it marks a return to the sensual pre-web realm (Web 0.0): Web 3.0 is "the biological, digital analogue web. Information is made of a plethora of digital values coalesced for sense and linked to the real world by analogue interfaces. . . . The web . . . is digested, thus integrated into our real world."[34] In this vision, Web 3.0 is a wet web, existing at the interface of the biological and the digital. Biologists and doctors have already begun to speculate about the possibilities of Web 3.0 for biomedicine, arguing that, "Whereas web 1.0 and 2.0 were embryonic, formative technologies, web 3.0 promises to be a more mature web where better 'pathways' for information retrieval will be created, and a greater capacity for cognitive processing of information will be built."[35] These individuals imagine an Internet like J. C. R. Licklider's "world brain"—a growing biological system. Biological practice and the problems and practices of Internetworking have become irreducibly connected. On the one hand, in the very near future one will not be able to do biology without using the tools of Web 3.0—vast amounts of heterogeneous data will be searchable and intelligible only by using the Semantic Web. On the other hand, the development of Web 3.0 has been driven by biological problems and biological data. Biology and networking are increasingly organized around a common set of practices and problems.

This periodization of the history of the web provides a suggestive analogy with the history of biology over the last seventy years. Biology 0.0 was the pre-informatic biology of the pre–World War II era—biologists understood function in morphological terms, as being determined by physical contact involving interlocking shapes (such as lock-and-key mechanisms). The rise of communications and information theory in the late 1940s and early 1950s inspired Biology 1.0, in which biological molecules now communicated with one another by exchanging information. Information was encoded in DNA, and it was this information that drove biological function. The rise of bioinformatics—from the mid-1980s onward—could be described as Biology 2.0. Biological objects could be represented and manipulated in the computer and circulated in digital spaces. Biology 2.0 entails all the kinds of changes and rearrangements of biological practice and biological knowledge that have been described in these chapters. But the analogy predicts a further shift—to Biology 3.0. We can already perceive the outlines of what this biology might look like. First, it draws on the tools of Web 3.0, particularly the Semantic Web, to create a hyperdata-driven biology. Not only will massive amounts of biological data be available online (this is already true), but these data may soon be semantically connected in such a way that discoveries about biological function can readily fall out of the data. But Biology 3.0 also constitutes an erasure of the boundary between the biological and the informatic: biological objects and their informatic representations will become apparently interchangeable. Data will be rich and reliable enough that doing a digital experiment by manipulating data will be considered the same thing as doing an experiment with cells and molecules.

Biology 3.0, like Web 3.0, is a return to a sensual, predigital biology because it refuses to recognize any difference between the biological and the digital, between "wet" and "informatic" biologies. The biological and the digital will become a single entity centered on the problems of organizing and sharing data. One might think of Biology 3.0 as an autopoietic machine, generating itself from a feedback loop between biology and information: the more deeply biology is conceived of as information, the more deeply it becomes a problem of data management, the more productive such an informatic conception becomes, and the more the notion of information is allowed to permeate the biological.[36] Biology 3.0 predicts the culmination of the process in which the biological and the informatic have become a single practice.

In other words, bioinformatics may disappear. The practice of using computers to generate or run simulations in physics is not designated separately from the rest of physics—there is no "phys-informatics." Such could be the case for bioinformatics—its practices seem likely to become so ubiquitous that it will be absorbed into biology itself. More precisely, what the notion of Biology 3.0 suggests is that the practices and knowledge associated with bioinformatics may gradually subsume those of the rest of biology. This process will not necessarily be a smooth one. As we have seen, bioinformatics has its opponents, particularly in those who believe that biology must proceed by asking and answering specific questions, by filling in the details of how organisms work, gene by gene, protein by protein. Others argue that computational work, detached from hard evidence from the lab bench, is pointless at best or misleading at worst. I am not suggesting that biology will completely lose contact with bodies and molecules. What this study suggests, however, is a trend toward the centralization of such "wet" work—at places such as the Broad Sequencing Center—and a decentralization of "dry" work. The "wet" work of biology may become increasingly confined to highly ordered and disciplined spaces designed to produce data with the greatest possible efficiency. Meanwhile, "dry" biology can be done anywhere, by anyone with a computer and an Internet connection.

Homo statisticus

Biology 3.0 also brings into focus the ways in which changes in biological practice may be tied to social and political changes. In the last forty years, biology has been transformed from an autonomous academic science into a venture that is intimately linked to health care, the pharmaceutical industry, and the speculative worlds of high capital. These changes have required exactly the sorts of modalities of work that bioinformatics has provided: ways of talking and working across disciplines, ways of globalizing biological knowledge, and ways of speeding up production and productivity. In other words, "bioinformatics" and "Biology 3.0" are ways of characterizing and describing large-scale shifts in the institutional context of biological work over the last decades.

We are just beginning to see how and where these far-reaching changes in the way biology is practiced may have consequences for human health, for our sociality and identity. The bioinformatic imperative for data gathering has led to several projects intended to vastly increase the amount of human sequence data. In September 2007, *PLoS Biology* published the complete diploid genome of J. Craig Venter,[37] which was followed closely by James Watson's genome in 2008.[38] Also launched in January 2008 was the 1000 Genomes Project, an international collaboration to completely sequence the genomes of several thousand anonymous participants in order to build an extensive catalog of human genetic variation.[39] Its aim, like that of its precursor, the HapMap project (completed in 2007, and which surveyed the genotypes—rather than the full genomes—of a large number of individuals), is to contribute to variation studies that link genetic variation to disease.[40] 1000 Genomes has now completed a pilot phase involving 1,092 individuals.[41] Over the next several years, the project will build a vast library of human sequence data (all publicly available) for bioinformatic work.

These public efforts are complemented by commercial enterprises in the field of personal genomics. Companies such as 23andMe are offering fee-based genotyping services for as little as $100. Complete genome or exome sequencing is approaching the consumer price range. George Church's Personal Genome Project offers a full-genome service to individuals willing to make their personal medical histories and full body photographs, as well as their genomes, publicly available.[42] All these projects aim to associate genotype with phenotype on the basis of statistics. In particular, genome-wide association studies (GWAS) are used to discover correlations between mutations at specific genomic loci and the presence of a phenotype or disease. The mechanism through which this locus might act to cause the trait is usually unknown—the statistical linkage is enough. A genetic profile from 23andMe, for instance, provides the user with information about their "risk factors" for over two hundred diseases, conditions, and traits (a few examples: celiac disease, rheumatoid arthritis, alcohol flush reaction, bipolar disorder, statin response, memory, male infertility).[43]

The rise of personal genomics indicates the extension of the bioinformatic paradigm into medicine. Just as our understanding of organisms and their functionality is increasingly dominated by statistical techniques applied to large data sets, our understanding of our own health, our bodies, and our futures is being similarly reconstructed. The body is no longer a collection of genes, but is rather becoming an accretion of statistics, a set of probabilities that refer to "risk factors" for various diseases and conditions. Paul Rabinow has argued that the genomic age creates new regimes of associations between individuals:

> The new genetics will cease to be a biological metaphor for modern society and will become instead a circulation network of identity terms and restriction loci, around which and through which a truly new type of autoproduction will emerge, which I call "biosociality." . . . Through the use of computers, individuals sharing certain traits or sets of traits can be grouped together in a way that . . . decontextualizes them from their social environment.[44]

Personal genomics is exactly this: a tool for tracking and computing risk and using those data to build not only new "socialities," but also new "individualities." This is *Homo statisticus*—a human constructed from the statistical residue of his or her genome.

In personal genomics, we are a collection of probabilities of having or acquiring particular traits; taking account of our own personal genome means living one's life according to such probabilities—exercising, eating, and seeking the appropriate medical interventions accordingly. "Everyone has something valuable," 23andMe's Anne Wojcicki assures us, "and that's your data."[45] This notion draws on the logic of bioinformatics, in which biomedical knowledge is made through the statistical analysis and comparison of sequence data. For Nikolas Rose, the important transformation here is the "creation of the person 'genetically at risk'"—the construction of an individual on the basis of sets of risk factors.[46] But there is a more fundamental transformation occurring here too: the creation of a person as a set of statistics and probabilities. Rose's "molecular optics" or "molecular biopolitics" is not just about how to live with, interact with, and deal with risk, but also about how to reimagine oneself as a statistical body, as *Homo statisticus*.[47]

Ulrich Beck has argued that contemporary society is best understood as a "risk society," in which the fundamental problem is not the distribution of "goods," but rather the management of "bads." In societies plagued by crime, impending environmental disaster, and other "manufactured" risks, the emphasis falls on how such risks might be mitigated and minimized. This mitigation is achieved through processes of "reflexive modernity," in which societies assess risk and take appropriate action toward decreasing it (for example, regulations on certain kinds of activities).[48] In personal genomics, this kind of risk thinking and risk discourse is adapted to individual bodies—these companies and projects trade on the notion that the risks of disease can be measured and minimized. Moreover, this thinking suggests a move toward "reflexive biology," in which individuals are able to measure and anticipate their level of risk and take appropriate action to mitigate it (appropriate diet, exercise, treatment, or testing). Reflexive biology is a set of practices oriented toward the calculation of risks in order to be able to intervene with preventative measures and treatments; the body is managed by computing and tracking a set of statistics.

A 2009 paper from the J. Craig Venter Institute analyzed the accuracy of some personal genomic services:

> Both companies report absolute risk, which is the probability that an individual will develop a disease. Absolute risk is derived from two parameters: "relative risk" and "average population disease risk." Relative risk is modelled from an individual's genetics. Average population disease risk varies depending on how one defines the population. For example, Navigenics distinguishes population disease risk between men and women

(for example, men are more likely to have heart attacks than women), whereas 23andMe primarily takes into account age (for example, incidence of rheumatoid arthritis increases with age). This ambiguity in the definition of a "population" underscores the caution one must exercise when interpreting absolute risk results.[49]

Although Navigenics and 23andMe are reporting risks, the crucial question is how such risks should be interpreted. Their interpretation relies on statistics—probabilities, averages, sampling, and so on. Understanding the body as a risky body entails first understanding it as a statistical one. Whereas the concept of "genetic risk" points toward genetic fatalism, the statistical body suggests something more open-ended—especially the possibility of using personal genomics to take control of one's own biological destiny. Moreover, the statistical body is not an individual body—based on a set of individualized risks—but a body that exists only in relation to other bodies (and in relation to a population). The statistical body carefully crafts "normalcy" (from intricately contrived mathematical and statistical analyses of genomes) and locates particular bodies in relation to that ideal. The original "human genome," reconstructed from the DNA of a few individuals and based on no data on human genetic variation, had very little power to claim that it represented an "ideal" or "normal" genome. With personal genomics, the 1000 Genomes Project, and other similar endeavors, the "normal" emerges clearly and powerfully from mathematical averages. Each one of us may soon be able to define ourselves statistically in relation to these means and to live out our lives accordingly.

Computers and networks have not only made this sort of work possible on a practical level, but have also generated a particular set of tools and practices and sensibilities for working with biological data. In personal genomics, we are just beginning to see how these tools and practices and sensibilities—which I have called bioinformatic—are manifesting themselves in health care. Problems of statistics, of managing massive amounts of data, of computing probabilities based on these data, are doing much more than opening up possibilities for preventative medicine. In the study of variation, the statistical body may come to describe a particular and powerful "normality" that is derived from probabilistic models that discover patterns among the vast amounts of disordered data from genomics and medical informatics. That ideal is likely to transform our own sociality—how we understand the similarities and differences between us and those around us—and our individuality—how we match ourselves up against that norm. The vast amounts of data produced by personal genomics would and will be meaningless without the tools of bioinformatic management and interpretation. Genomes would be nothing but a long, disordered string of As, Gs, Ts, and Cs—sequence "out of sequence." But the statistical techniques of bioinformatics allow biologists to bring order to these data, to reconstruct "out of sequence" an object that has biological, medical and social meaning.

Acknowledgments

The archival research and fieldwork for this project were supported by two grants. A grant from the National Science Foundation (0724699) allowed me to travel freely to conduct interviews and collect archival material. Likewise, much of my ethnographic fieldwork was supported by a grant from the Wenner-Gren Foundation for Anthropological Research. My writing was supported by grants from Harvard University in 2009–2010 and Nanyang Technological University in 2011–2012.

The richest sources of data for this project were my conversations with those who spent their days (and often nights) working in bioinformatics. It is my first priority to thank the seventy-five individuals who agreed to interviews. Their contributions remain anonymous, but nevertheless invaluable. The anthropological components of this work would not have been possible without the generous efforts of all those who organized my visits, answered my innumerable questions, helped me to learn, gave me space (and computers) in their labs, and generally tried hard to understand what on earth a historian was doing taking notes at their lab meetings. In the Department of Biology at the Massachusetts Institute of Technology, thanks must first of all go to Chris Burge, who trusted me enough to let me work in his lab, to give me a "real" computational project, and to let me sit in on his first-year graduate computational biology seminar. Without this beginning, the project might have ended up much worse off. The Burge lab proved a welcoming environment, and many of my informants became my friends.

My visit to the Harvard Institute of Proteomics (then housed within the Broad Institute) was hosted by Joshua LaBaer. It was Yanhui Hu who allowed me to occupy half her desk, showed me the ropes, gave me a project, took me to seminars, introduced me to people, and supported my work at every turn. Also at HIP, thanks to Jason Kramer (who put up with me sharing his office for three and a half months), Catherine Cormier, Janice Williamson, Dongmei Zuo, and Sebastian Degot. My visit to the European Bioinformatics Institute was organized by Ewan Birney and Cath Brooksbank, who worked hard to make sure my visit was productive. Paul Flicek volunteered to host my visit and made sure that I had space to work and access to computer systems and meetings. I worked most closely with Stefan Gräf, who deserves special credit for taking time out of his extremely busy schedule to teach me how the Ensembl database worked.

Locating and collecting archival materials also depended on the help of many individuals. At Stanford, Will Snow and Henry Lowood gave me valuable advice on who to talk to and where to begin my research at the University Archives. Lowood also made available to me an online collection of material belonging to Edward Feigenbaum and archived through the Self-Archiving Legacy Tool (SALT). I also made extensive use of an online collection of archival materials compiled by Timothy Lenoir and Casey Reas. At the National Center for Biotechnology Information, Dennis Benson allowed me to peruse a rich collection of materials on the history of GenBank and the NCBI that had come into his possession. At the European Bioinformatics Institute, Graham Cameron permitted me to spend several weeks examining his personal files, which provided a mine of historical documents on GenBank, EMBL-Bank, and the EBI itself. Michael Ashburner similarly allowed me access to his files on EMBL-Bank, EBI, and the Gene Ontology. Staff at the American Philosophical Society Archives in Philadelphia, the National Library of Medicine in Bethesda, and the Stanford University Archives provided valuable help in accessing collections belonging to Walter Goad, Douglas Brutlag, and Edward Feigenbaum.

Getting this work into its final form depended on the help and support of many colleagues. My friends in the Department of History of Science at Harvard offered deep intellectual engagement as well as moral support. Aaron Mauck, Grischa Metlay, Alex Csiszar, Daniela Helbig,

Tal Arbel, Funke Sangodeyi, Alex Wellerstein, Nasser Zakariya, Lukas Rieppel, and Alistair Sponsel—it was a pleasure to share my graduate student experience with all of them. I also owe thanks to many other science studies scholars who have, at various times, read, listened, and offered comments on parts of the manuscript. Michael Fischer deserves special mention for reading and commenting on an early version of the whole manuscript. I am also deeply grateful for the support of my new colleagues in the School of Humanities and Social Sciences at Nanyang Technological University, Singapore.

Karen Darling, my editor at University of Chicago Press, has encouraged and nurtured this work over several years. My thanks go to her and to the Press's reviewers, as well as my copyeditor Norma Roche, all of who helped to improve the book immeasurably.

My academic mentors deserve pride of place on this list. They gave me the freedom to roam wide interdisciplinary spaces and the discipline and rigor necessary to weave the project together into a whole. Sarah Jansen always made sure my work was grounded in a sound knowledge of the history of biology. Stefan Helmreich tirelessly read and commented on my work, and his ideas and suggestions led me in directions I would never have otherwise discovered. Peter Galison supported and encouraged my somewhat unusual project from beginning to end. At Harvard, I was also challenged by Sheila Jasanoff, whose introductory class on science, power, and politics sharpened my disciplinary tools. At MIT, a class on ethnography led by Michael Fisher and Joe Dumit served as a valuable introduction to the anthropology of science. Hannah Landecker and Chris Kelty, who visited Harvard in 2007–2008, offered me much friendly encouragement and advice. Finally, both as an undergraduate and as a graduate student, I was lucky enough to share both the intellectual energy and the friendship of two remarkable mentors, Sam Schweber and Steven Shapin—both of these individuals have inspired and enriched me as an academic and as a person.

Work on this book has spanned six years and ranged across four continents. None of this would have been possible without an incredible amount of support from those closest to me: my parents, Gillian and Peter, my sister, Lara, and my wife, Yvonne. I count myself extraordinarily lucky to have them.

Archival Sources

The following archival sources were consulted for this project. The full citations of the collections are given here. An abbreviated citation ("WBG papers," etc.) is given in both the notes and the bibliography.

EAF papers—Edward A. Feigenbaum papers (Stanford University, Department of Special Collections and University Archives, SC340).

GC papers—Graham Cameron personal files.

GenBank papers—GenBank historical documents (National Center for Biotechnology Information, National Library of Medicine). Made available by permission of Dennis Benson.

JAL papers—Joshua Lederberg papers (Modern Manuscripts Collection, History of Medicine Division, National Library of Medicine, MS C 552).

WBG papers—Walter Goad papers (American Philosophical Society, MS Coll 114).

Notes

INTRODUCTION

1. Dyson, "Our Biotech Future."

2. A selection of readings: Forman, "Behind Quantum Electronics"; Kaiser, "Postwar Suburbanization"; Kaiser, "Scientific Manpower"; Galison, *Image and Logic*.

3. Glaser, "The Bubble Chamber."

4. Eisenhower, "Farewell."

5. Weinberg, *Reflections*, 40.

6. Weinberg, "Criteria." Elsewhere, Weinberg spoke of the "triple diseases" of Big Science, "journalitis, moneyitis, and administratisis" (Weinberg, "Impact of Large Scale Science"). Another important critic of Big Science was the historian and physicist Derek J. de Solla Price; see Price, *Little Science, Big Science*. A good historical summary of the debates is provided in Capshew and Rader, "Big Science."

7. Anderson, "More Is Different."

8. At least in terms of funding, this "debate" appeared to be quite one-sided—with the high-energy physicists far out in front—until the end of the Cold War; as Daniel Kevles has pointed out, the demise of the Superconducting Super Collider left physicists in the 1990s searching for new ways to justify their work. See Kevles, "Big Science and Big Politics."

9. Latour and Woolgar, *Laboratory Life*, 15ff.

10. Gilbert, "Towards a Paradigm Shift."

11. This account is partly based on an early draft of Gilbert's paper from 1990 that was examined by Fujimura and Fortun. See Fujimura and Fortun, "Constructing Knowledge."

12. Gilbert, "Towards a Paradigm Shift."

13. Gilbert, "Towards a Paradigm Shift."

14. Barnes and Dupré, *Genomes*. This view is also informed by the work of Brian Cantwell Smith on problems of ontology in computer science. Smith, *On the Origin of Objects*.

15. Although recently, increasing recognition of the value of data is changing this. For example, the journal *Gigascience* (launched in 2012 by the Beijing Genomics Institute) will accept and host large-scale life science data sets in its database. The aim remains, however, to promote and publish "studies" of these data sets.

16. *Nature*, "Community Cleverness Required." This editorial appeared as part of a *Nature* special issue on "big data."

17. Some representative examples from sociology and history: Appadurai, *Social Life of Things*; Schlereth, *Material Culture*; Bourdieu, *Outline*: Miller, *Material Culture*. Some representative examples from the history of science: Galison, *Image and Logic*; Kohler, *Lords of the Fly*; Chadarevian and Hopwood, *Models*; Rasmussen, *Picture Control*; Rader, *Making Mice*; Creager, *Life of a Virus*; Turkle, *Evocative Objects*.

18. Fortun, "Care of the Data."

19. See Kohler, *Lords of the Fly*.

20. During 2007 and 2008, I conducted participant observation–style fieldwork at three locations: at the laboratory of Christopher Burge, in the Department of Biology at the Massachusetts Institute of Technology, Cambridge, Massachusetts; at the Broad Institute, Cambridge, Massachusetts; and at the European Bioinformatics Institute, Hinxton, United Kingdom.

21. This is a good point at which to distinguish this account from Joe November's very important historical accounts of the first uses of computers in biology (November, "Digitizing Life," and November, *Biomedical Computing*). November's narrative begins with the first computers developed in the wake of World War II and concludes with the application of artificial intelligence to biology at Stanford in the 1960s and 1970s. Here, I begin—for the most part—in the 1970s and attempt to come as close to the present as possible. In a sense, this account can be seen as a sequel to November's: even at the end of the 1970s, computers were extremely rare in biology and medicine, and those who used them remained far from the mainstream. To understand the ubiquitous, seemingly universal role that computers have come to play today, we need to understand what has happened with computers and biology *since* the 1970s.

22. Writing a history of bioinformatics is in some ways automatically an ahistorical project: no one did "bioinformatics" in the 1960s or 1970s. Of course, the sets of practices that became bioinformatics later on can be traced to practices from those earlier decades, but the "founding" individuals were funded by different sources, organized in different disciplines, published in different journals, and understood their work as having significantly different aims. By tracing back these strands, I necessarily exclude or gloss over many of the other ways in which computers were used to do biological work from the 1960s onward. Bioinformatics is centered almost entirely on a relatively narrow set of computation problems focused on DNA, RNA, and protein

sequences. These problems have become by far the most visible applications of computers in biology. But computers were, and are, used for a range of other things (numerical simulations of population dynamics, simulation and prediction of protein folding, visualization of molecules, experimental design, simulation of evolution, simulation of the brain, instrument control, etc.). This book is not a history of all the ways in which computers have been used in biology. Rather, its purpose is to understand why a certain set of problems and practices came to be picked out and to dominate our understanding of life.

23. Approaches that focus on "dematerialization" in various forms include Mackenzie, "Bringing Sequences to Life"; Pottage, "Too Much Ownership"; Waldby, *Visible Human Project*; and Parry, *Trading the Genome*. Others (Lenoir, "Science and the Academy," and Thacker, *Biomedia*) describe bioinformatics as a shift in the "media" of biological work.

24. It is important to distinguish this argument from arguments about gene-centrism that have been put forward by numerous scholars (notably including Keller, *Century of the Gene*, and Nelkin and Lindee, *The DNA Mystique*). Those arguments make the case that molecular biology has been dominated by a near-dogmatic obsession with genes and other associated regulatory elements rather than epigenetic or cytoplasmic aspects of biology; DNA acts as a kind of "genetic program" or "master molecule" that controls all aspects of an organism's behavior. I agree with these assessments. However, my claim is slightly different in that it does not involve the claim that it is genes that are in control; rather, the sequence as a whole is the object of analysis. The sequence could be RNA, DNA, or protein, acting alone or in combination, with no molecule necessarily at the top of the hierarchy. All sequences, not just genes or DNA, contain meaning and importance for biological function that might be understood by performing statistical analyses.

CHAPTER ONE

1. November, *Biomedical Computing*, 7.

2. November, *Biomedical Computing*, 8.

3. November argues that earlier efforts in biomedical computing laid the groundwork for the large-scale computerization of biology during and after the 1980s: "Today's life scientists are only able to take advantage of computers because they have access to tools and methods developed during a costly, ambitious, and largely forgotten attempt to computerize biology and medicine undertaken by the NIH in the late 1950s and early 1960s" (November, *Biomedical Computing*, 8). But as November goes on to say, using these new tools and methods involved finding ways of making biology more quantitative and excluding nondigital aspects of biology. That is, biologists were developing data-oriented forms of biological work. I would add only that, until the advent of sequence, this data-oriented biology was practiced by only a small minority of biologists.

4. Campbell-Kelly and Aspray, *Computer*; Ceruzzi, *Reckoners*; Ceruzzi, *History of Modern Computing*; Edwards, *Closed World*.

5. Edwards, *Closed World*, 43.

6. For a detailed account of Turing's life and work, see Hodges, *Alan Turing*.

7. Campbell-Kelly and Aspray, *Computer*, 73–74.

8. Campbell-Kelly and Aspray, *Computer*, 80–83.

9. Edwards, *Closed World*, 51.

10. Edwards, *Closed World*, 65–66.

11. The ENIAC was designed to compute firing tables, which required the solution of differential equations (see Polacheck, "Before the ENIAC"). It was eventually used in calculating implosion for the hydrogen bomb, which also involved numerical solution of partial differential equations.

12. Galison, *Image and Logic*, 490.

13. Galison, *Image and Logic*, 492.

14. Ceruzzi, *History of Modern Computing*, 30.

15. On OR, see Fortun and Schweber, "Scientists and the Legacy."

16. For more on operations research, see Kirby, *Operational Research*.

17. On the relationship between computers and operations research, see Agar, *Government Machine*; Wilkes, "Presidential Address."

18. It was very difficult and time-consuming to program early computers. This meant that sections of programs (subroutines) were often reused for different purposes. It may have been possible to imagine creating an entirely new program from scratch, but for many purposes this would have taken far too long. Furthermore, in most cases, it was the same small group of individuals who were responsible for taking a problem and "programming" it into the machine; these people evolved unique styles of problem solving and machine use.

19. Perhaps the most revealing use of computers in biology before 1960 was in the attempts to "crack" the genetic code. Since the genetic code, like the computer, was understood as an information processing system, the computer was seen by some as the ideal tool with which to attack the problem. These attempts failed. Kay, *Who Wrote the Book of Life?*

20. November, "Digitizing Life," 170.

21. Ledley, "Digital Electronic Computers."

22. Ledley, *Use of Computers*, x.

23. Ledley, *Use of Computers*, xi.

24. Ledley, *Use of Computers*, 1–12.

25. See also the 1965 volume by Stacy and Waxman, *Computers in Biomedical Research*. These authors concluded that "the pattern of development which is emerging in the biomedical sciences appears to be similar in almost all respects to the scientific evolution of the physical sciences" (3).

26. November, *Biomedical Computing*, 270.

27. On the development of minicomputers, see Ceruzzi, *History of Modern Computing*, 127–135.

28. Lenoir, "Shaping Biomedicine," 33–35.

29. Friedland et al., "MOLGEN" [EAF papers].

30. Friedland, "Knowledge-Based Experimental Design," iv–v.

31. The spread of computers into labs during the 1980s no doubt had much to do with the increasing storage and speed and decreasing costs associ-

ated with the personal computer. A good example of the way scientists began to use personal computers in the 1980s can be found in *Nature*, "An Eye for Success."

32. Bishop, "Software Club."

33. This is not to say computers were not used for anything else. Biologists continued to use computers for all the kinds of activities listed in this paragraph. Simulation and prediction of protein folding, which has played a particularly visible role, emerged in the 1990s as a distinct area of research within bioinformatics. This is also a case in which sequence data (here, amino acid rather than DNA or RNA sequence) play a crucial role; for bioinformatics, problems of structure prediction can be reduced to problems of collecting and analyzing data in order to model the relationship between amino acid sequence and three-dimensional spatial data.

34. The thesis is Goad, "A Theoretical Study of Extensive Cosmic Ray Air Showers" [WBG papers]. Other work is cited below.

35. On computers at Los Alamos, see Metropolis, "Los Alamos Experience"; Anderson, "Metropolis, Monte Carlo, and the MANIAC"; Metropolis and Nelson, "Early Computing at Los Alamos."

36. Mackenzie, "The Influence of the Los Alamos," 189.

37. See, for instance, Goad, "Wat" [WBG papers].

38. See, for instance, Goad and Johnson, "A Montecarlo Method."

39. Goad, "A Theoretical Study of Extensive Cosmic Ray Air Showers" [WBG papers].

40. Goad and Cann, "Theory of Moving Boundary Electrophoresis"; Goad and Cann, "Theory of Zone Electrophoresis"; Goad and Cann, "Theory of Transport."

41. This is not a complete account of Goad's work during this period. He also contributed to studies on the distribution of disease, the digestion of polyoxynucleotides, and protein–nucleic acid association rates, among others. Some of this work involved other kinds of transport phenomena, and all of it involved statistical methods and computers.

42. Goad, "Vita." [WBG papers]

43. Goad, "T-division" [WBG papers].

44. On the difficulties of theoretical biology, see Keller, "Untimely Births of Mathematical Biology," in *Making Sense of Life*, 79–112.

45. On Dayhoff, see Strasser, "Collecting, Comparing, and Computing Sequences."

46. Los Alamos National Laboratory, "T-10 Theoretical Biology and Biophysics" [WBG papers].

47. See Dietrich, "Paradox and Persuasion"; Hagen, "Naturalists."

48. Goad (undated letter to Japan[?]) [WBG papers].

49. Ulam, "Some Ideas and Prospects"; Goad, "Sequence Analysis."

50. For a history of sequence comparison algorithms, see Stevens, "Coding Sequences." Some seminal papers from this group: Beyer et al., "A Molecular Sequence Metric"; Waterman et al., "Some Biological Sequence Metrics"; Smith and Waterman, "Identification of Common Molecular Subsequences."

51. On T-10 culture, see Waterman, "Skiing the Sun."

52. Theoretical Biology and Biophysics Group, "Proposal to Establish a National Center" [WBG papers].

53. Los Alamos National Laboratory, "$2 Million Earmarked" [WBG papers].

54. George I. Bell, letter to George Cahill [WBG papers].

55. For more, see Strasser, "Collecting, Comparing, and Computing Sequences."

56. Strasser, "GenBank"; Strasser, "The Experimenter's Museum."

57. Those clearly identifiable by name and institution (others used a non-obvious username): Bob Sege (Yale University), Hugo Martinez (UCSF), Howard Goodman (UCSF), Perry Nisen (Albert Einstein), Jake Maizel (NIH), Bill Pearson (Johns Hopkins), Brian Fristensky (Cornell), Hans Lehrach (EMBL Heidelberg, Germany), Allen Place (Johns Hopkins), Schroeder (University of Wisconsin), Frederick Blattner (University of Wisconsin), Rod Brown (Utah State University), David W. Mount (University of Arizona), Fotis Kafatos (Harvard), Jean-Pierre Dumas (CNRS, Paris), Dusko Ehrlich (Université de Paris), Dan Davison (SUNY at Stony Brook), Jeffrey S. Haemer (University of Colorado), Tom Gingeras (Cold Spring Harbor), Paul Rothenberg (SUNY at Stony Brook), Tom Kelly (Johns Hopkins), Clyde Hutchison (University of North Carolina), Allan Maxam (Harvard Medical School), Robin Gutell (UCSC), Mike Kuehn (SUNY at Stony Brook), John Abelson (UCSD), Margaret Dayhoff (National Biomedical Research Foundation), Mark Ptashne (Harvard), Irwan Tessman (Purdue), Brad Kosiba (Brandeis), Rick Firtel (UCSD), Walter Goad (Los Alamos Scientific Laboratories), R. M. Schwartz (National Biomedical Research Foundation), H. R. Chen (Atlas of Proteins), Stephen Barnes (Washington University), Andrew Taylor (University of Oregon), Gerschenfeld (Stanford), Annette Roth (Johns Hopkins), Mel Simon (UCSD), Mark Boguski (University of Washington Medical School), Kathleen Triman (University of Oregon), and Stoner (Brandeis). National Institutes of Health, "MOLGEN Report," 17-18 [EAF papers].

58. A word is in order about Ostell's name. James Michael Ostell was born James Michael Pustell, and it is this name that appears on some of his earlier published papers. He changed it sometime around 1986, and Ostell is the name that appears on his PhD thesis.

59. Ostell, "A Hobby Becomes a Life."

60. CP/M (Control Program for Microcomputers) is an operating system developed by Gary Kildall in 1973-1974.

61. Ostell "A Hobby Becomes a Life."

62. Ostell, "Evolution," 3.

63. Ostell, "Evolution," 3-4.

64. Pustell and Kafatos, "A Convenient and Adaptable Package of DNA Sequence Analysis Programs for Microcomputers."

65. Pustell and Kafatos, "A Convenient and Adaptable Package of DNA Sequence Analysis Programs for Microcomputers," 51-52.

66. Pustell and Kafatos, "A Convenient and Adaptable Package of Com-

puter Programs for DNA and Protein Sequence Management, Analysis and Homology Determination," 643.

67. Interview with Jim Ostell, April 7, 2008, Bethesda, Maryland.

68. See Pustell and Kafatos, "A Convenient and Adaptable Microcomputer Environment for DNA and Protein Sequence Manipulation and Analysis."

69. Wong et al., "Coding and Potential Regulatory Sequences"; Baumlein et al., "The Legumin Gene Family."

70. Ostell, "Evolution," chapter 4, 23.

71. The story of Ostell's work at the NCBI is continued, in the context of database design, in chapter 5.

72. Ostell, "Evolution," chapter 11, 4–5.

73. Ostell, "Evolution," chapter 11, 58.

74. Ostell, "Evolution," introduction and thesis summary, 15.

75. For Hogeweg's usage, see Hogeweg, "Simulating."

76. Searching in PubMed for "bioinformatics" automatically includes searches for "computational biology" and "computational" AND "biology."

77. Hunter, "Artificial Intelligence and Molecular Biology."

78. Hunter, quoted in International Society for Computational Biology, "History of ISCB."

79. This conference is usually taken by bioinformaticians and computational biologists to be the first in the field. The annual ISMB conferences are still the most widely attended and most widely respected conferences in bioinformatics.

80. Stephan and Black, "Hiring Patterns," 6–7.

81. Bolt, Beranek, and Newman, "BBN Progress Report," 2 [GenBank papers].

82. Beynon, "CABIOS Editorial."

83. Vilain, e-mail to Feigenbaum [EAF papers].

84. Olsen, "A Time to Sequence," 395.

85. For instance, see Aldhous, "Managing the Genome Data Deluge"; Roos, "Bioinformatics."

86. The independence of the HGP from GenBank and the NCBI is also evidenced by the fact that it set up its own sequence database, the Genome Database. This database was based at Johns Hopkins University and funded by the Howard Hughes Medical Institute. See Cook-Deegan, *Gene Wars*, 290–291.

87. US Congress, House of Representatives, "To establish the National Center for Biotechnology Information." For a fuller description of the legislative history of the NCBI, see Smith, "Laws, Leaders, and Legends."

88. "The European Bioinformatics Institute (EBI)," 2–4. [GC papers]

89. Davison et al., "Whither Computational Biology," 1.

90. An earlier contender for this honor is Lesk, *Computational Molecular Biology*. It is perhaps the first significant attempt to generate a synoptic overview of the field, but it remains a collection of papers rather than a true textbook. A number of other textbooks arrived in the next three years: Baldi and Brunak, *Bioinformatics*; Bishop and Rawlings, *DNA and Protein Sequence*; Durbin et al., *Biological Sequence Analysis*; Gusfield, *Algorithms on*

Strings; Setubal and Medianas, *Introduction to Computational Molecular Biology*.

91. Waterman, *Introduction*.

92. Altman, "Editorial," 549.

93. Altman, "Editorial," 550.

94. Expectation maximization is a statistical method; Monte Carlo methods come from physics, as we have seen; simulated annealing is a probabilistic simulation; dynamic programming was originally invented in the context of systems analysis; cluster analysis is a kind of statistical data analysis; neural networks are simulations based on the brain; genetic algorithms are simulations based on evolution; Bayesian inference is a statistical method based on Bayes's theorem; stochastic context-free grammars are probabilistic models of behavior.

95. Taylor, "Bioinformatics: Jobs Galore."

96. Black and Stephan, "Bioinformatics: Recent Trends," i.

97. *Nature*, "Post-genomic Cultures."

98. *Nature*, "Post-genomic Cultures."

99. Butler, "Are You Ready for the Revolution?," 758.

100. Roos, "Bioinformatics," 1260

101. Butler, "Are You Ready for the Revolution?," 760.

102. Knight, "Bridging the Culture Gap," 244.

CHAPTER TWO

1. These controversies can also be seen in disagreement over the origins of bioinformatics: should it be dated to the first attempts at systematically organizing biological data, the first paper databases, the first computer databases, the first networked databases, or the first algorithms for data analysis?

2. Usually the word "data" refers to *any* set of numbers or facts that are operated on in order to draw an inference or conclusion, regardless of context. Such a definition is inappropriate here because bioinformatics is so tightly bound to computational practices that make data what they are for their biological users.

3. For this project, I made no attempt to gain direct access to corporate spaces. However, over the course of many interviews with current and former private-sector employees, I managed to build up a reasonably detailed picture of work in corporate bioinformatics. There are also a few companies that are exclusively devoted to bioinformatics (for example, IntelliGenetics and Panther Informatics).

4. Throughout the book, first names appearing alone are pseudonyms. Where first and last names are given, this indicates the use of an individual's real name.

5. For example, the paper announcing the sequencing of the honeybee in 2006 had over 300 authors from 90 institutions. Indeed, authorship had to be broken down into overall project leadership (2), principal investigators (2), community coordination (7), annotation section leaders (12), caste development and reproduction (14), EST sequencing (8), brain and behavior (42),

development and metabolism (35), comparative and evolutionary analysis (6), funding agency management (4), physical and genetic mapping (12), ribosomal RNA genes and related retrotransposable elements (4), gene prediction and consensus set (20), honeybee disease and immunity (18), BAC/fosmid library construction and analysis (9), G+C content (5), transposable elements (2), gene regulation (30), superscaffold assembly (6), data management (14), chromosome structure (6), population genetics and SNPs (13), genome assembly (7), (A+T)-rich DNA generation (6), tiling arrays (6), anti-xenobiotic defense mechanisms (11), and DNA sequencing (26). Honeybee Genome Sequencing Consortium, "Insights into Social Insects." Of course, the physics community has been used to working in large interdisciplinary groups since the first large-scale particle accelerator experiments in the 1950s. Some biologists I spoke to said that biology was just "catching up" with physics in terms of its attitudes toward management and interdisciplinarity.

6. Interview with Charles DeLisi, February 12, 2008, Boston Massachusetts.

7. Interview with Richard J. Roberts, February 25, 2008, Ipswich, Massachusetts.

8. Very recently a third group has begun to emerge from dedicated bioinformatics and computational biology PhD programs, like the CSB (Computational and Systems Biology) program at MIT. Many of these individuals have had significant training in and exposure to *both* computer science and biology.

9. Those trained in biology would be more inclined to publish in *Cell* or the *Journal of Molecular Biology*, while those with training in other fields might look to *Bioinformatics, Journal of Computational Biology*, or even *Machine Learning*.

10. Burge encouraged his students to spend time doing both wet-lab and dry-lab work, often giving them problems that would require them to engage in both. However, he made sure that students spent a minimum of three months on one side before switching; in his experience, any less would lead to less than satisfactory work.

11. This description is based on a paper from the Burge lab: Yeo et al., "Identification and Analysis of Alternative Splicing Events." The supplementary material contains a detailed description of the experimental procedures.

12. This is a description (based on the interview) of the project undertaken by the computer science graduate student working in the Burge lab, as described above.

13. According to Burks, as of 2008, approximately 2×10^{11} bases had been sequenced. The genomes of one hundred representatives of all species would amount to 3×10^{18} bases, or about 10 million times more sequence than the 2008 figure.

14. Stark et al., "Discovery of Functional Elements."

15. For a recent example of this kind of argument, see Stein, "The Case for Cloud Computing." In fact, Charles DeLisi first pointed out this possibility in 1988: DeLisi, "Computers in Molecular Biology."

16. For instance, the cover headline for the BioIT World issue of April 2008 was "Weathering the Next-Gen Data Deluge." The story it referred to

was Davies, "DNA Data Deluge." During the time of my fieldwork, the most attention was given to the Solexa-Illumina (San Diego, CA) and 454 Life Sciences (Branford, CT) GS FLX sequencing machines. Solexa machines can produce up to 6 billion bases per two-day run (www.illumina.com), while 454 claims 100 million bases per seven-hour run (www.454.com). The Broad Institute was also beginning to test prototypes of the new SOLiD machines from Applied Biosystems and the "open source" machines designed by George Church's lab at Harvard Medical School. In addition, Helicos Biosciences (Kendall Square, Cambridge, MA) was preparing to launch an even more powerful machine capable of sequencing from a single DNA molecule (Harris et al., "Single Molecule DNA Sequencing").

17. During the time I spent in the lab at MIT, Burge was in the process of trying to convince the MIT biology department to purchase a Solexa machine to be shared between several labs.

18. A detailed review of the alternative splicing literature can be found in Black, "Mechanisms of Alternative Pre-messenger RNA Splicing."

19. For examples of work that asks such questions, see Wang et al., "Alternative Isoform Regulation," and Sandberg et al., "Proliferating Cells." In the classical molecular genetics approach, the regulation of a specific gene is studied, and the findings from that case may or may not generalize to other genes. The computational approach necessarily picks out those aspects of regulation that are universal, or at least relatively common. A clear example is the study of microRNA targets. Early studies of microRNAs focused on a couple of specific examples: *lin-4* and *let-7*. These microRNAs often had target sites that lacked perfect complementarity to the 5′ end of the microRNA (the "seed"), obscuring the importance of this region. This problem was quickly corrected by computational studies. See Lewis et al., "Prediction of Mammalian MicroRNA Targets," and Rajewsky, "MicroRNA Target Prediction."

20. For example, see Fischer, "Towards Quantitative Biology." This review gives examples of how bioinformatic knowledge is expected to be able to generate quantitative predictions about diseases and interaction of drugs with the body.

21. Fortun, "Projecting Speed Genomics." "Speed genomics" refers to biology done on a large and expensive scale, in big, centralized labs, with expensive machines.

22. Fortun, "Projecting Speed Genomics," 30.

23. See chapter 3 for a more detailed discussion of notions of the regimes of "productivity" associated with computing.

24. The chimpanzee paper is fairly typical of the kind of large-scale work that the Broad undertakes. Its website lists active areas of research as "deciphering all the information encoded in the human genome; understanding human genetic variation and its role in disease; compiling a complete molecular description of human cancers . . ." In addition to many genome projects, its work has included attempts to completely characterize human genetic variation (HapMap, 1000 Genomes), to completely characterize cancer and its genetics, and to create an RNAi library that covers every known human gene.

25. The Solexa machines allowed the direct sequencing of mRNA taken from cells. Older methods usually required mRNA to be reverse-transcribed into DNA before sequencing. This was usually done in short fragments known as expressed sequence tags (ESTs).

26. Previous studies include Nigro et al., "Scrambled Exons"; Cocquerelle et al., "Splicing with Inverted Order of Exons"; Capel et al., "Circular Transcripts"; Zaphiropoulos, "Circular RNAs"; Caudevilla, "Natural Transsplicing"; Frantz et al., "Exon Repetition in mRNA"; Rigatti et al., "Exon Repetition."

27. One bioinformatic study of exon scrambling had been performed using ESTs, concluding that scrambling events were either nonexistent or extremely rare: Shao et al., "Bioinformatic Analysis of Exon Repetition."

28. Compare my description with Galison's description of the discovery of neutral currents in the 1970s: Galison, *How Experiments End*, 188–198.

29. Fieldwork notes, August 9, 2007.

30. Of course, it is literally possible to go through a program line by line to understand its behavior. However, this is almost never how it is done. Rather, you debug a program by running it on some real data, seeing what happens, seeing whether the results make sense. If not, you use the error to find the part of the program that needs attention, alter it, and try again.

31. In my work, the Solexa machines themselves were regions of ignorance—for proprietary reasons, Illumina did not reveal all the details of how their sequencing process worked, and hence the characteristics and reliability of the data it produced were largely unknown. The lab did extensive computational analysis on the data it received to try to find the extent to which it could rely on those data.

32. See Popper, *Logic of Scientific Discovery*.

33. Bacon, *New Organon*, 110.

34. On "discovery science," see Aebersold et al., "Equipping Scientists," and Boogerd et al., *Systems Biology*.

35. Allen, "Bioinformatics and Discovery." For further discussion of hypothesis-driven research, see Wiley, "Hypothesis-Free?"; Gillies, "Popper and Computer Induction"; Allen, "Hypothesis, Induction-Driven Research and Background Knowledge"; Smalheiser, "Informatics and Hypothesis-Driven Research."

36. This debate has been carried on vigorously in the blogosphere. See, for instance, Young, "Hypothesis-Free Research?"

37. "Proposals to base everything on the genome sequence by annotating it with additional data will only increase its opacity." Brenner, "Sequence and Consequences."

38. Allen, "In Silico Veritas," 542.

39. In later chapters, we will detail more precisely the methods used for data management and data analysis in bioinformatics.

40. Garcia-Sancho, in particular, has argued for the importance of specific computing architectures in shifting the practices and epistemologies of nematode worm biology (see Garcia-Sancho, "From the Genetic to the

Computer Program"). But other recent scholarship seems to downplay both the novelty of "big data" and the distinction between "hypothesis-driven" and "hypothesis-free" biology (see Leonelli, "Introduction").

41. DeLisi, "Computers in Molecular Biology," 47.

42. Franklin, "Exploratory Experiments."

43. Brown and Botstein, "Exploring the New World."

44. On the increasing importance of hypothesis-free biology, see Kell and Oliver, "Here Is the Evidence."

45. Strasser, "Data-Driven Sciences."

46. Georges Cuvier, Louis Agassiz, and Charles Darwin, for instance, all dealt with large numbers of objects and paper records. Calling all this "data," however, seems to impose a present-day category on historical actors. The *Oxford English Dictionary* offers two definitions of "data": one based on philosophy (facts forming the basis of reasoning) and one from computing (symbols on which operations are performed). Suggesting that natural history deals with "data" runs the risk of confusing or conflating these two definitions.

47. Bacon, "The New Atlantis."

CHAPTER THREE

1. One way to appreciate the special value of sequence data is to examine the Bermuda rules. In 1996 and 1997, a set of meetings among the genomics community established rules for how and when data were to be shared and who was entitled to use them. These rules recognized the unique epistemic status of data (as something less than knowledge) and attempted to prevent the direct exchange of data for financial value.

2. Davis, "Sequencing the Human Genome," 121. Davis compared the HGP with Nixon's ill-fated "war on cancer" of the 1970s.

3. Davis, "Sequencing the Human Genome," 21.

4. Shapin, "House of Experiment."

5. At the start of my fieldwork, it was usually my practice to ask informants and interviewees who else among their friends, colleagues, and professional acquaintances I should talk to. However, the separation of the two groups made it such that, having begun talking to computational biologists, it was difficult to gain an introduction to a bioinformatician.

6. One bioinformatics blog, written by a graduate student, half-jokingly divided bioinformatics into six different "career paths": Linux virtuoso ("the LV performs all their research at the command line: vi edited bash scripts chained together using shell pipes"), early adopter ("always working on the latest area of research, system biology synthetic biology, personal genomics"), old school ("blinkered to change in tools and technology, the Old School is doing their analysis in Fortran on a Windows 95 Pentium II"), data miner ("their everyday tools are mixed effect regression, hidden Markov models, and the fearsome neural gas algorithm"), perfect coder ("produces code like poetry and, after a five second glance, even your dog knows what the script does"), wet-lab bioinformatician ("while others have their heads in the clouds thinking about theories and algorithms, the WB is getting his hands dirty with real data as it is

being produced"). By analogy with a fantasy role-playing game (called World of Bioinformatics Quest), pursuers of these careers have different attributes for coding, statistics, presentation/writing, research focus, and collaboration. "Getting the right attributes is therefore critical, and playing to your strength will result in more Papers™ and Grants™." Barton, "World of Bioinformatics Quest." This post provides a detailed, if amusing, description of the various sorts of practices in bioinformatics and how biologists perceive the differences between them.

7. An epitope tag is a region of a folded protein that is recognized by an antibody. Tagging a protein with an epitope allows it to be recognized by antibodies, which allow the construction of a simple test for the presence of a protein.

8. Broad Institute, "Who is Broad?"

9. This observation invites comparison with Shapin's arguments about seventeenth-century laboratory spaces in which certain technical practices were hidden from view (Shapin, "Invisible Technician").

10. Goffman, *Presentation of Self*, chapter 3.

11. Goffman gives the example of a watch-repair shop in which the watch is taken into a back room to be worked on: "It is presented to [the customer] in good working order, an order that incidentally conceals the amount and kind of work that had to be done, the number of mistakes that were first made before getting it fixed, and other details the client would have to know before being able to judge the reasonableness of the fee that is asked of him." Goffman, *Presentation of Self*, 117.

12. Goffman, *Presentation of Self*, 125. He adds that the back region is often reserved for "technical" standards while the front is for "expressive" ones (126).

13. For several months while I was working in the MIT biology department, and regularly interviewing scientists across the road at 7CC, I remained completely unaware of the existence of 320 Charles.

14. For more on the design of the Broad Institute, and especially its "transparency," see Higginbotham, "Collaborative Venture"; Higginbotham, "Biomedical Facility"; Silverberg, "The Glass Lab."

15. Broad Institute, "Philanthropists."

16. Crohn's disease is a genetic disorder affecting the intestinal system. On the Broad's work on Crohn's disease, see Rioux et al., "Genome-Wide Association Study."

17. Gieryn, "Two Faces," 424.

18. Silverberg, "The Glass Lab."

19. Womack et al., *Machine*, 12–13.

20. Womack et al., *Machine*, 57.

21. Womack et al., *Machine*, 129.

22. Womack et al., *Machine*, 152.

23. Broad Institute, "Rob Nicol."

24. Six Sigma is a business management strategy first implemented by Motorola that attempts to quantify and control variation in output by carefully monitoring and correcting product defects. The name reflects the aim to imple-

ment processes that produce products that are defective only 0.00034% of the time; that is, in which defects are normally distributed, but occur only as rarely as events six standard deviations from the mean (6-sigma events). See Stamatis, *Six Sigma Fundamentals*.

25. Vokoun, "Operations Capability Improvement," 6.

26. Chang, "Control and Optimization."

27. Chang, "Control and Optimization," 65.

28. Vokoun, "Operations Capability Improvement."

29. Vokoun, "Operations Capability Improvement," 51.

30. Vokoun, "Operations Capability Improvement," 49–50.

31. Vokoun, "Operations Capability Improvement," 55–56.

32. Vokoun, "Operations Capability Improvement," 105.

33. Several other Sloan students also applied lean production principles and other management techniques to aspects of the Broad: Scott Rosenberg analyzed the computer finishing process (Rosenberg, "Managing a Data Analysis Production Line"), and Kazunori Maruyama studied the electrophoretic sequencing process itself (Maruyama, "Genome Sequencing Technology").

34. Disproportionate with respect to the Broad Institute as a whole, and with respect to the profession of "biologists" as a whole. The Broad is particularly proud of its large community of Tibetans, most of whom work at the sequencing center; this occasioned a visit by the Dalai Lama to the Broad Sequencing Center in 2003 during his visit to MIT. The pipette used by His Holiness is still mounted on the wall of the sequencing center, together with his portrait.

35. Rosenberg, "Managing a Data Analysis Production Line," 75.

36. Rosenberg, "Managing a Data Analysis Production Line," 82–83.

37. Rosenberg, "Managing a Data Analysis Production Line," 59–71.

38. Rosenberg, "Managing a Data Analysis Production Line," 78.

39. Vokoun, "Operations Capability Improvement," 69–70.

40. National Human Genome Research Institute, "Genome Sequencing Centers (U54)."

41. Agar, *Government Machine*. On the Treasury, see chapter 8.

42. Campbell-Kelly and Aspray, *Computer*, 105.

43. Haigh, "Chromium-Plated Tabulator."

44. Cerruzi, *History*, 32–33.

45. Cortada, *Information Technology*, 160. On the history of computers in business, see also Edwards, "Making History."

46. The Beijing Genomics Institute is an especially interesting example, since it, unlike the Broad, places its sequencing activities front and center. Rather than suggesting a fundamentally different model, though, the BGI may be revealing of the global dynamics of science: China and the BGI itself may be said to constitute a "back space" that produces materials to be consumed in the "front spaces" of US and European biomedical institutions. The world's largest manufacturer is also emerging as the major manufacturer of sequence.

47. This kind of view is also prevalent in synthetic biology, in which attempts have been made to strip organisms down to the bare essentials

necessary for life (the so-called "minimum genome project"). See Glass et al., "Essential Genes."

48. Edward Yoxen's notion of "life as a productive force" (Yoxen, "Life as a Productive Force"), Marilyn Strathern's view of "nature, enterprised-up" (Strathern, *After Nature*), Catherine Waldby's "biovalue" (Waldby, *Visible Human Project*), Sunder Rajan's "biocapital" (Rajan, *Biocapital*), and Nicolas Rose's "bioeconomics" (Rose, *Politics of Life Itself*) all seek to capture the essence of what is going on between biology, biotechnology, medicine, politics, and the economy. For more on the relationship between biocapital and the production of sequence, see Stevens, "On the Means of Bio-production."

CHAPTER FOUR

1. The science studies literature has emphasized the importance of physical and geographic spaces: see Collins, "TEA Set"; Shapin, "Pump and Circumstance." Very little attention has been given to virtual spaces and proximities, however. Recently Collins has asked whether "electronic communication makes any difference to the nature of expertise." (He concludes that it does not: see Collins, "Does Electronic Communication Make Any Difference.")

2. Vokoun, "Operations Capability Improvement," 28–30.

3. Chang, "Control and Optimization," 12–14.

4. Person, "Operational Streamlining," 24.

5. Person, "Operational Streamlining," 28–29.

6. See www.phrap.com. See also Ewing et al., "Base Calling."

7. Rosenberg, "Managing a Data Analysis Production Line," 16–17.

8. In fact, any sequence that is worked on during a given day is submitted or resubmitted to GenBank at the end of that day; only finished sequences acquire "finished" status in the database, however.

9. This point is made in detail in Barnes and Dupré, *Genomes*, 103–109.

10. For a more detailed account of this transition, see Stevens, "Coding Sequences."

11. Dowell et al., "Distributed Annotation System."

12. Smith, "Ontology," 155.

13. Smith, "Ontology," 160.

14. Smith, "Ontology," 162.

15. Lewis, "Gene Ontology," 104.

16. Lewis, "Gene Ontology," 104.

17. Lewis, "Gene Ontology," 104; interview with Michael Ashburner, December 10, 2008, Cambridge, UK.

18. Ashburner, "On the Representation of 'Gene Function.'" Ashburner suggested linking GO to Stanford's "Ontolingua" project (http://ksl-web .stanford.edu/knowledge-sharing/ontolingua/) and Schulze-Kremer's ontology for molecular biology (Schulze-Kremer, "Ontologies for Molecular Biology").

19. Interview with Michael Ashburner, December 10, 2008, Cambridge, UK.

20. Technically, GO is structured as a set of "directed acyclic graphs"—like

a tree hierarchy, but where each term can have multiple parents. For instance, if a particular gene was involved in making hexose, its GO terms would include "hexose biosynthetic process," "hexose metabolic process" (since any biosynthetic process is [less specifically] a metabolic one), and "monosaccharide biosynthetic process" (since hexose is [less specifically] a monosaccharide). Terms can be semantically linked to one another through relationships such as "is_a" or "regulates_positively." GO conforms to the specifications of a Web Ontology Language (OWL) since its designers expect biological databases to be searched over the web (see the Conclusion for more details on the relationship between bio-ontologies and web ontologies).

21. Gene Ontology Consortium, "Gene Ontology," 25.

22. "The Consortium emphasized that GO was not a dictated standard. Rather, it was envisioned that groups would join the project because they understood its value and believed it would be in their interest to commit to the ontology. Participation through submission of terminology for new areas and challenging current representation of functionality was encouraged." Bada, "Short Study."

23. Interview with Midori Harris, November 26, 2008, Hinxton, UK. In fact, there are four full-time editors and roughly forty further contributing researchers who have permission to write to (that is, edit) the GO Concurrent Versioning System.

24. OBO Foundry, "Open Biological and Biomedical Ontologies."

25. One such ontology is called the Basic Formal Ontology.

26. Smith, "What Is an Ontology?"

27. See http://www.flickr.com/groups/folksonomy/. See also Ledford, "Molecular Biology Gets Wikified."

28. Smith, "What Is an Ontology?"

29. Smith, "What Is an Ontology?"

30. Smith, "Ontology as the Core Discipline."

31. Leonelli, "Centralizing Labels."

32. For instance, the DAVID (Database for Annotation, Visualization and Integrated Discovery) web resource allows users to combine ontology terms and other identifiers from many different sources. See http://david.abcc.ncifcrf.gov/ and Sherman et al., "DAVID Knowledgebase." The plausibility of GO becomes particularly important in high-throughput experiments in which tens or even hundreds of genes may be simultaneously up- or down-regulated by some exogenous conditioning (a drug, for example). The lab biologist then wishes to find out whether these genes have anything in common. This can be done by testing for "enrichment" of annotation terms among the set of genes. For example, if 50 of the 60 up-regulated genes were annotated with the GO term "cell-cycle," the biologist might be able to conclude that the drug has an effect on the cell cycle.

33. On the Semantic Web and data sharing in biology, see Ure et al., "Aligning Technical and Human Infrastructures."

34. Field et al., "Towards a Richer Description."

35. Again, how the machines are physically linked together in real space was of usually of little importance (unless something went wrong). Machines

could be physically linked in complicated ways via routers, switches, virtual
private networks, login hosts, load balancers, servers, and cache machines.

36. Sanger Centre, "The Farm FAQ."

37. European Bioinformatics Institute, "Ensembl Release Coordination."

38. In 2011, Ensembl was accessed by about 3.4 million unique web (IP)
addresses per year (http://www.ebi.ac.uk/Information/News/press-releases/
press-release-28112011-directors.html). A quick search of PubMed yields
about 450 articles that refer to Ensembl.

39. Turner, "Video Tip."

40. In the discussion of databases in chapter 5, I examine how specific
organizational structures affect the content of biological knowledge. Here I
am arguing that some sort of organization, some sense of spatial thinking, is
necessary for granting "knowledge" status to pieces of biological data in the
first place.

41. The notion of interconnectedness comes from Tomlinson,
Globalization.

42. For more on how knowledge travels, see Morgan and Howlett, *How
Well Do Facts Travel?*

CHAPTER FIVE

1. Bruno Strasser argues that biological databases are an instance of the
"natural history tradition" within modern biology (Strasser, "Collecting and
Experimenting"; Strasser, "GenBank"). The "moral economies" of collecting, exchanging, describing, comparing, and naming natural objects, Strasser
argues, belong more to the "wonder cabinet, the botanical garden, or the
zoological museum" than the laboratory. The biggest problem databases
had to solve, Strasser continues, were the social problems associated with
"collection of data, scientific credit, authorship, and the intellectual value of
collections."

2. Bowker and Star, *Sorting Things Out*.

3. By suggesting that databases work as schemes of classification, I am suggesting their importance as tools for dividing and ordering the world in ways
similar to schemes for classifying and ordering living things, diseases, and races
(see Koerner, *Linnaeus*, and Suárez-Díaz and Anaya-Muñoz, "History, Objectivity," on the importance of plant and animal classification; Rosenberg and
Golden, *Framing Disease*, on the classification of disease; and Duster, "Race
and Reification," on race classifications).

4. Most of the historical and sociological literature that touches on biological databases has focused on their role as "communication regimes." Stephen Hilgartner, in particular, has argued that the genome databases are "novel
tools for scientific communication" (Hilgartner, "Biomolecular Databases").
Although databases do provide efficient means of distributing and sharing biological data and have initiated new standards and expectations for data access,
it is not sufficient to understand biological databases as passive technologies
for storing and redistributing information, akin to scientific journals. (On data
access policies, see Hilgartner, "Data Access Policy.")

5. Haigh, "Veritable Bucket." On SAGE, see also Edwards, *Closed World*, chapter 3.

6. See Bachman's 1973 Turing Award acceptance speech (Bachmann, "Programmer as Navigator").

7. Codd, "A Relational Model," 377.

8. Codd called the process of distributing data in this way "normalization."

9. Haigh, "Veritable Bucket," 43.

10. Hunt, "Margaret Oakley Dayhoff." See also Dayhoff and Kimball, "Punched Card Calculation."

11. Ledley was invited by George Gamow to join the RNA Tie Club in 1954. See chapter 1 of this book for more on Ledley's use of computers in biology, and for a detailed account of Ledley's career, see November, "Digitizing Life," chapter 1.

12. November, "Digitizing Life," 73–74.

13. November argues that this was one of two visions for computerizing biomedicine in the 1950s and 1960s. The other, advocated by Howard Aiken, was a mathematization of the life sciences on the model of physics. November, "Digitizing Life," 169–170.

14. Ledley, *Use of Computers*, 12.

15. Dayhoff and Ledley, "Comprotein," 262.

16. Dayhoff, "Computer Aids."

17. See Pauling and Zuckerkandl, "Divergence and Convergence," and Zuckerkandl and Pauling, "Molecules as Documents."

18. Margaret O. Dayhoff to Carl Berkley, February 27, 1967. Quoted in Strasser, "Collecting and Experimenting," 111.

19. Dayhoff and Eck, *Atlas*.

20. Dayhoff and Eck, *Atlas*, viii.

21. Dayhoff and Eck, *Atlas*, 36.

22. To find superfamily relationships, Dayhoff compared sequences both with each other and with randomly generated sequences; related sequences showed a much higher match score than could be generated by chance. Superfamilies organized proteins into groups according to evolutionary relationships; through this method, many proteins within the same organisms (even those with very different functions) could be shown to have similarities to one another. Dayhoff et al., "Evolution of Sequences."

23. Dayhoff, "Origin and Evolution."

24. Dayhoff, "Computer Analysis."

25. Strasser, "Collecting and Experimenting," 112.

26. Strasser, "Collecting and Experimenting," 113–114.

27. John T. Edsall to Dayhoff, November 4, 1969. Quoted in Strasser, "Collecting and Experimenting," 116.

28. Dayhoff to Joshua Lederberg, draft, March 1964. Quoted in Strasser, "Collecting and Experimenting," 117.

29. For more details, see Strasser, "Experimenter's Museum."

30. Smith, "History of the Genetic Sequence Databases."

31. Smith, "History of the Genetic Sequence Databases," 703. Strasser also gives an account of the history of GenBank, in which he rehearses his theme

that sequence databases belong to the "natural history tradition." In particular, he portrays GenBank's success (in contrast with Dayhoff's "failure") as a result of its commitment to "open," as opposed to proprietary, science. Strasser, "GenBank."

32. For more on protein crystallography databases, see Berol, "Living Materials."

33. Smith, "History of the Genetic Sequence Databases," 702.

34. Anderson et al., "Preliminary Report" [GenBank papers].

35. National Institutes of Health, "Report and Recommendation," 1 [GC papers].

36. Ohtsubo and Davison, letter to Jordan, 1 [GenBank papers]. In addition to this letter from the Department of Microbiology at the State University of New York at Stony Brook, Jordan and Cassman received letters supporting the database from Mark Pearson at the Frederick Cancer Research Center, Andrew Taylor at the Institute of Molecular Biology at the University of Oregon, and Charles Yanofsky at the Department of Biological Sciences at Stanford.

37. Theoretical Biology and Biophysics Group, "Proposal to Establish a National Center" [WBG papers].

38. Theoretical Biology and Biophysics Group, "Proposal to Establish a National Center" [WBG papers]

39. In fact, it was a fully developed relational database management system.

40. Smith, "History of the Genetic Sequence Databases," 704.

41. Blattner, "Second Workshop," 3 [GenBank papers].

42. National Institutes of Health, "Agenda: 3rd Workshop" [GenBank papers].

43. Many internal NIH documents from this period reiterate the justifications for funding the database. In particular, see National Institutes of Health, "Funding Plan" [JAL papers].

44. Jordan, "Request for Contract Action," 1 [JAL papers].

45. As part of the Department of Energy, and consequently as an agent of the government, Los Alamos was prohibited from competing directly for a federal contract from the NIH. The arrangement was that BBN became the prime contractor, subcontracting the collection efforts to Goad's team. The double bid from Los Alamos arose because, again as a government agent, the lab could not be seen to be favoring one private firm over another and thus had to agree to collaborate with multiple companies.

46. Smith, "History of the Genetic Sequence Databases," 705.

47. Keller, Century of the Gene, 51–55.

48. The first successful attempts to isolate the causative genes for specific diseases were made in the 1980s. For example, in 1989, Francis Collins and his co-workers successfully isolated a single point mutation that caused cystic fibrosis. Rommens et al., "Identification of the Cystic Fibrosis Gene."

49. The work of Russell Doolittle provides an exemplar of this gene-centric practice. In 1983, Doolittle, using his own customized protein sequence database, discovered a surprising similarity between an oncogene and a growth hormone. Doolittle's discovery has become a folkloric tale illustrating the

power and importance of nucleotide sequence databases. The original paper is Doolittle et al., "Simian Sarcoma Virus." See also Doolittle, "Some Reflections."

50. For a contemporary review of work being done with sequences, see Gingeras and Roberts, "Steps Toward Computer Analysis."

51. For instance, in the sample table given here, the exon runs from the 273rd nucleotide to the 286th nucleotide of the listed sequence, and the TATA box from the 577th to the 595th.

52. Admittedly, this quote is hardly from an unbiased source, since it formed part of the NBRF's protest against the awarding of the GenBank contract to BBN and Los Alamos. National Biomedical Research Foundation, "Rebuttal" [GenBank papers].

53. Schneider, "Suggestions for GenBank" [WBG papers].

54. Los Alamos National Laboratory, "Los Alamos Progress Report," 1 [GenBank papers].

55. Kabat, "Minutes" [WBG papers].

56. Fickett and Burks, "Development of a Database," 37 [GenBank papers].

57. This struggle had become quite public knowledge by 1986. See Lewin, "DNA Databases."

58. National Institutes of Health, "Proposed Statement of Work" [GenBank papers].

59. The discussion of the contractor selection can be found in Kirschstein and Cassatt, "Source Selection," 2 [GenBank papers].

60. Robert Cook-Deegan traces the origins further to a meeting in 1984. See Cook-Deegan, "Alta Summit."

61. Department of Energy, Office of Health and Environmental Research, "Sequencing the Human Genome," 4.

62. National Center for Human Genome Research, "Options for Management of Informatics" [GenBank papers]; Jordan, "Record" [GenBank papers].

63. Bolt, Beranek, and Newman, "BBN Progress Report," 2 [GenBank papers].

64. Cinkosky et al., "Technical Overview," 1 [GenBank papers].

65. Cinkosky et al., "Technical Overview," 2–3 [GenBank papers].

66. Cinkosky and Fickett, "A Relational Architecture" [GenBank papers].

67. Los Alamos National Laboratory, "Database Working Group Meeting" [GenBank papers].

68. Gilbert, "Towards a Paradigm Shift."

69. From a review article: Bork et al., "Predicting Function."

70. A large number of examples can be cited: Mushegian and Koonin, "Gene Order"; Mushegian and Koonin, "Minimal Gene Set"; R. Himmelreich et al., "Comparative Analysis"; Koonin et al., "Comparison of Archaeal and Bacterial Genomes"; Dandekar et al., "Conservation of Gene Order"; Huynen et al., "Genomics."

71. National Institutes of Health, National Library of Medicine, "Long Range Plan," 27.

72. National Institutes of Health, National Library of Medicine, "Talking One Genetic Language."

73. For the initial hearing, see US Congress, House of Representatives, "Biotechnology."

74. US Congress, Office of Technology Assessment, "Mapping Our Genes," 98–99. A more comprehensive account of the politicking surrounding the establishment of the NCBI can be found in Smith, "Laws, Leaders, and Legends."

75. US Congress, House of Representatives, "Biotechnology," 39.

76. Interview with David Lipman, April 9, 2008, Bethesda Maryland.

77. Jordan, letter to Director, Office of Human Genome Research [GenBank papers].

78. National Institute of General Medical Science and National Library of Medicine, "First Status Report" [GenBank papers].

79. For more on Ostell's background, see chapter 1.

80. Dubuisson, *ASN.1*.

81. Ostell, "Integrated Access," 730.

82. Ostell, "Integrated Access," 731.

83. Ostell, "Integrated Access," 731.

84. Ostell et al., "NCBI Data Model," 19–20.

85. Ostell et al., "NCBI Data Model," 20.

86. Ostell et al., "NCBI Data Model," 43.

87. Indeed, the output of the NCBI Basic Local Alignment Search Tool (BLAST) became a Bioseq.

88. Fischer, "Towards Quantitative Biology."

89. Galperin and Cochrane, "Nucleic Acids Research."

90. Bowker and Star, *Sorting Things Out*, 27.

91. Examples include Thompson, *On Growth and Form*; Waddington, ed., *Towards a Theoretical Biology*; Kaufmann, *Origins of Order*; Prusinkiewicz and Lindenmeyer, *Algorithmic Beauty of Plants*; Bertalanffy, *General Systems Theory*; Varela et al., "Autopoiesis." Artificial life also considered itself a kind of theoretical biology; see Helmreich, *Silicon Second Nature*, and Keller, *Making Sense of Life*, chapter 9.

92. Keller, *Making Sense of Life*, 237.

CHAPTER SIX

1. Visualization and problems of representation have played a major part in recent studies of science and medicine. See, for instance, Hannah Landecker's study of the role of microcinematography (Landecker, "Microcinematography") or Peter Galison's study of images in high-energy physics (Galison, *Image and Logic*). The role of computers in image making has also been examined by Norton Wise (Wise, "Making Visible").

2. Daston and Galison have much to say about the role of images in recent science (Daston and Galison, *Objectivity*). They argue that recent scientific image making marks a shift from representation to *presentation*—from attempts to faithfully represent nature to efforts to "do things with images." Taking examples primarily from nanofabrication, they contend that image generation has become inseparable from the engineering of nanoscale objects. These "haptic images" function "as a tweezer, hammer, or anvil of nature: a tool to make

things change" (383). Nanoengineers are not using visualization as a tool for finding some sort of "truth to nature," but as a tool to make nature. What is going on in bioinformatics is different—biologists are not trying to construct or manipulate external objects using images. Bioinformatic images are not attempts to create "objective" representations of biology, but rather attempts to address the problem of managing the vast amounts of data and numbers of computations that make it practically impossible to fully "check up" on a computer's result. Images are manipulated not as means of manipulating nature, but as means of analyzing and understanding it.

3. The work of Lily Kay (*Who Wrote the Book of Life?*) and Evelyn Fox Keller (*Making Sense of Life*) has sensitized historians to the importance of textual metaphors in molecular biology. But other nontextual metaphors are now becoming increasingly important. For one interesting take, see the work of Natasha Myers on kinaesthetic metaphors in biology (Myers, "Modeling Proteins").

4. The same is true in other biological fields such as artificial life (see Helmreich, *Silicon Second Nature*).

5. The MAC in Project MAC originally stood for Mathematics and Computation; it was later renamed Multiple Access Computer, then Machine Aided Cognitions or Man and Computer.

6. Levinthal et al., "Computer Graphics."

7. Francoeur and Segal, "From Model Kits." See also Francoeur, "The Forgotten Tool," and Francoeur, "Cyrus Levinthal."

8. Ledley, *Use of Computers.*

9. On the significance of scientific atlases, see Daston and Galison, "Image of Objectivity." Daston and Galison argue that atlases became means to mechanically and objectively represent and order the phenomena of nature.

10. See Stevens, "Coding Sequences."

11. The first description of this method can be found in Gibbs and McIntyre, "The Diagram." For a detailed history of dot plots, see Wilkinson, "Dot Plots."

12. See Stevens, "Coding Sequences," and Needleman and Wunsch, "A General Method."

13. For instance, a bioinformatics textbook describes the algorithm in visual language: "The best alignment is found by finding the highest-scoring position in the graph, and then tracing back through the graph through the path that generated the highest-scoring positions." Mount, *Bioinformatics*, 12.

14. To explain in a little more detail, Needleman-Wunsch works by identifying small matching regions of nucleotides (or residues in the case of proteins) and then trying to extend those alignments outward. Working from left to right, the grid is populated with scores according to either finding matches or inserting gaps in the sequence. When all the squares in the grid have been computed, the highest score appears somewhere at the bottom right; it is then possible to read-off the best alignment from right to left by "tracing back" what sequence of moves resulted in this high score.

15. National Academy of Sciences et al., *Information Technology*, 1.

16. McCormick et al., "Visualization," 7.

17. Friedhoff and Benzon, *Visualization*.

18. Sulston et al., "Software for Genome Mapping."

19. Durbin and Thierry-Mieg, "ACEDB."

20. Durbin and Thierry-Mieg, "ACEDB."

21. Durbin and Thierry-Mieg, "ACEDB."

22. Indeed, object-oriented programming languages were invented for the purpose of physical modeling. On the need for a general theory of ontology when using an object-oriented programming language, see Smith, *On the Origin of Objects*. Smith argues that "computer scientists wrestle not just with notions of computation itself, but with deeper questions of how to understand the ontology of worlds in which their systems are embedded. This is so not for the elementary (if significant) reason that anyone building a functioning system must understand the context in which it will be deployed, but for the additional one that computational systems in general, being intentional, will represent at least some aspects of those contexts. The representational nature of computation implies something very strong: that *it is not just the ontology of computation that is at stake; it is the nature of ontology itself*" (42, Smith's emphasis).

23. Dahl and Nygaard, *Simula*, 1–2.

24. Dahl and Nygaard, *Simula*, 4.

25. AceDB, "Introduction to AceDB."

26. Interview with Ewan Birney, November 12, 2008, Hinxton, UK.

27. Hubbard et al., "Ensembl Genome Database Project," 38. For more on Birney's career, see Hopkin, "Bring Me Your Genomes," and Birney, "An interview."

28. Stabenau et al., "Ensembl Core."

29. Stabenau et al., "Ensembl Core," 930.

30. Stabenau et al., "Ensembl Core," 931.

31. Stabenau et al., "Ensembl Core," 931.

32. Hubbard et al., "Ensembl Genome Database Project," 39.

33. For more on the internal workings of the Ensembl website, see Stalker et al., "Ensembl Web Site."

34. The website can be seen at http://www.ensembl.org.

35. Kent et al., "Human Genome Browser," 996.

36. Kent et al., "Human Genome Browser," 1004.

37. For more on autoSql, see Kent and Brumbaugh, "autoSql and autoXml."

38. Smith, *On the Origin of Objects*, 49. Smith's emphasis.

39. Kent et al., "Human Genome Browser," 996.

40. For a detailed description of using the UCSC Genome Browser, see Harte and Diekhans, "Fishing for Genes."

41. Harte and Diekhans, "Fishing for Genes," 3.

42. Maps of various sorts have played an important role in genetics and genomics. See the significant work on mapping cultures in Rheinberger and Gaudillière, *From Molecular Genetics to Genomics*, and Gaudillière and Rheinberger, *Classical Genetic Research*.

43. Burge and Karlin, "Prediction of Complete Gene Structures."

44. To give a sense of their ubiquity, Weinstein estimates that there were about 4,000 scientific papers published between 1998 and 2008 that included heat maps. Moreover, the introduction of the use of heat maps for microarray analysis (Eisen et al., "Cluster Analysis," discussed below) was, in July 2008, the third most cited article in *Proceedings of the National Academy of Sciences USA*. See Weinstein, "Postgenomic Visual Icon."

45. The simplest way to do this is to use a distance metric based on the Pythagorean theorem extended to higher-dimensional spaces. For instance, if one measured the levels of four proteins in four patients, the "distance" between the measured levels in any two patients would be given by

$$\sqrt{(a_i - a_j)^2 + (b_i - b_j)^2 + (c_i - c_j)^2 + (d_i - d_j)^2},$$

where i and j are the two patients and a, b, c, d are the four proteins. Once all the distance pairs have been measured in this way, the algorithm would cluster the two patients with the smallest distance into a node on the tree. The position of the node is given by an average of i and j. The process is then repeated using the remaining two patients and the new node, again searching for the shortest distance. The process is repeated until all the patients are joined into a single tree.

46. Weinstein reports that in one case, a data set from molecular pharmacology was reduced from a 8.3-meter scroll to a one-page heat map with 4,000 lines. Weinstein, "Postgenomic Visual Icon," 1773.

47. Eisen et al., "Cluster Analysis."

48. Sneath, "The Application of Computers."

49. Hagen, "The Introduction of Computers."

50. Wilkinson and Friendly, "History of the Cluster Heat Map." The Paris data are taken from Loua, *Atlas statistique*.

51. Weinstein, "Postgenomic Visual Icon," 1772.

52. Alternative splicing of mRNA transcripts is described in chapter 2. For a detailed but introductory account of alternative splicing, see Matlin et al., "Understanding Alternative Splicing."

53. The amount of data produced was equivalent to about four complete human genomes. The production of these data was made possible by the sequencing speed of the new machines such as those made by Illumina. See www.illumina.com and Bentley et al., "Accurate Whole Human Genome Sequencing."

54. Wang et al., "Alternative Isoform Regulation."

55. On "mechanical objectivity," see Daston and Galison, *Objectivity*.

56. Fry, "Computational Information Design."

57. Fry, "Computational Information Design," 11.

58. See http://www. benfry.com/salaryper/.

59. To see many of these tools in action, go to http://benfry.com/projects/.

60. Fry, "Computational Information Design," 82–83.

61. Fry, "Computational Information Design," 89–90.

62. Fry, "Computational Information Design," 110.

63. Reas and Fry, *Processing*. See also www.processing.org.

64. Casey Reas, quoted in Fry, "Computational Information Design," 126.

65. Agar, "What Difference Did Computers Make?"

66. "Boundary objects are objects which are both plastic enough to adapt to local needs and constraints of the several parties employing them, yet robust enough to maintain a common identity across sites. . . . They have different meanings in different social worlds but their structure is common enough to more than one world to make them recognizable means of translation" (Star and Griesemer, "Institutional Ecology"). Here I am concerned not so much with mediation of different social worlds as with mediation of different logical regimes; pictures are "boundary objects" that are generated from internal (often database) representations of biological data but can also be interpreted as biologically meaningful.

67. For example, Lenoir, "Science and the Academy"; November, "Digitizing Life," iv.

68. It is of interest here to compare this argument with those by James C. Scott and others about the role of visualization in the realms of governance and power. Scott argues that certain mappings require a narrowing of vision that "brings into sharp focus certain limited aspects of an otherwise far more complex and unwieldy reality" in a manner that permits "knowledge, control, and manipulation." Scott, *Seeing Like a State*, 11. Visualization and representation techniques in bioinformatics can also be understood as ways of taming, controlling, and ultimately remaking the spaces and territories of sequences.

69. Smith, *On the Origin of Objects*, 16.

CONCLUSION

1. Goetz, "Craig Venter."

2. Human Microbiome Project Consortium, "Structure, Function"; Human Microbiome Project Consortium, "A Framework."

3. Gilbert et al., "Meeting Report."

4. This effect is beginning to be examined in the literature; see, for example, Davies, *$1000 Genome*.

5. For examples of biologists making this connection, see Rothberg and Leamon, "Development and Impact of 454 Sequencing"; Mardis, "Impact of Next-Generation Sequencing Technology."

6. For technical details, see Rothberg et al., "Integrated Semiconductor Device." A video describing the sequencing process can be found at http://www .iontorrent.com/the-simplest-sequencing-chemistry/.

7. Wojcicki, "Deleterious Me."

8. Most obviously, it raises a range of privacy issues concerning the use of individual genetic data. In addition, this research would not be peer-reviewed or subjected to institutional review through the usual channels. This fact poses additional concerns about the possible uses of this research by 23andMe and its customers.

9. Ewan Birney, "An Interview." A similar story was related to me by Fran Lewitter, who was recruited from BBN to the Brandeis biology department in the late 1980s in order to upgrade its network connectivity. Interview with Fran Lewitter, December 17, 2007, Cambridge, Massachusetts.

10. I use the past tense here because, although cgi.pm was the standard for a long time, it has now been largely eclipsed by other methods, including those using JSP (Java Server Pages) and PHP (PHP Hypertext Processor).

11. Stewart, "An Interview with Lincoln Stein."

12. Quoted in O'Reilly et al., "The Importance of Perl."

13. Stein, "How Perl Saved the Human Genome Project."

14. Berners-Lee, "The Web."

15. Berners-Lee, "Transcript of Talk."

16. Berners-Lee, "Digital Future."

17. World Wide Web Consortium, "RDF Primer."

18. World Wide Web Consortium, "OWL."

19. Smith et al., "OWL."

20. OWL evolved in part from a language known as DAML-OIL (DARPA Agent Markup Language—Ontology Inference Layer) that was funded by ARPA and managed by Bolt, Beranek, and Newman.

21. For lists of biomedical ontologies, see the website of the National Center for Biomedical Ontology (http://www.bioontology.org/ncbo/faces/pages/ontology_list.xhtml) and the website of the Open Biomedical Ontologies (http://obofoundry.org/).

22. Feigenbaum et al., "The Semantic Web."

23. Berners-Lee convinced his research group at MIT that biology, because of the richness of its data, was the best test case for the Semantic Web. Another example of an attempt to use Semantic Web technology for biology is the Interoperable Informatics Infrastructure Consortium (I3C). I3C aimed to create tools for data exchange and data management by developing common protocols and software. One of its first creations was the Life Sciences Identifier (LSID), which provided a standard way to refer to and access biological information from many different kinds of sources on the web.

24. Neumann, "Semantic Web."

25. Neumann, "Semantic Web."

26. Small-world networks are those in which the average distance between any two nodes is much less than that for a random network of the same size. Scale-free networks are those for which the distribution of the number of edges over vertices follows a power law. This means that the networks have hubs—some few very important nodes to which many other nodes are connected—but many nodes with one or very few edges.

27. Barabási, and Oltvai, "Network Biology," 106.

28. Han et al., "Effect of Sampling."

29. Kelleher, "Google, Sergey, and 23andMe."

30. Goetz, "23andMe."

31. Cisco and Intel (Dossia) also have plans to roll out medical records systems. See Lohr, "Microsoft Rolls Out Personal Health Records." BBN has a Semantic Web system for health care; Accelrys offers life science–specific modeling, simulation, and text mining applications.

32. For example, the collaborative filtering sites del.icio.us and Flickr.

33. Wells, "Web 3.0 and SEO."

34. Zand, "Web 3.0."

35. Giustini, "Web 3.0 and Medicine."

36. Maturana and Varela define an autopoietic machine as "a machine organized as a network of processes of production of components that produce the components which . . . through their interactions and transformations continuously generate and realize the network of processes and relations that produced them." (Maturana and Varela, "Autopoiesis," 78–79).

37. Levy et al., "Diploid Genome Sequence."

38. Wheeler et al., "Complete Genome of an Individual."

39. See http://www.1000genomes.org/.

40. Genotyping involves sequencing selected portions of an individual's genome; these portions are usually those known to be highly variable among individuals.

41. The 1000 Genomes Project Consortium, "An Integrated Map."

42. See http://www.personalgenomes.org/. The project began by sequencing ten individuals, but Church eventually hopes to enroll a hundred thousand participants.

43. See https://www.23andMe.com/health/all/.

44. Rabinow, "Artificiality and Enlightenment," 241–243.

45. Wojcicki, "Deleterious Me."

46. Novas and Rose, "Genetic Risk."

47. On molecular biopolitics, see Rose, *Politics of Life Itself*, 11–15.

48. Beck, *Risk Society*.

49. Ng et al., "Agenda for Personalized Medicine."

Bibliography

AceDB. "Introduction to AceDB." Accessed April 15, 2009. http://www.acedb.org/introduction.shtml.

Aebersold, Ruedi et al. "Equipping Scientists for the New Biology." *Nature Biotechnology* 18, no. 4 (April 2000): 359.

Agar, Jon. *The Government Machine: A Revolutionary History of the Computer.* Cambridge, MA: MIT Press, 2003.

———. "What Difference Did Computers Make?" *Social Studies of Science* 36, no. 6 (2006): 869–907.

Aldhous, Peter. "Managing the Genome Data Deluge." *Science* 262, no. 5133 (1993): 502–503.

Allen, John F. "Bioinformatics and Discovery: Induction Beckons Again." *BioEssays* 23, no. 1 (2001): 104–107.

———. "Hypothesis, Induction-Driven Research and Background Knowledge: Data Do Not Speak for Themselves. Replies to Donald A. Gillies, Lawrence A. Kelley, and Michael Scott." *BioEssays* 23, no. 9 (2001): 861–862.

———. "In Silico Veritas—Data-Mining and Automated Discovery: The Truth Is In There." *EMBO Reports* 2, no. 7 (2001): 542–544.

Altman, Russ B. "Editorial: A Curriculum for Bioinformatics: The Time Is Ripe." *Bioinformatics* 14, no. 7 (1998): 549–550.

Anderson, Carl et al. "Preliminary Report and Recommendations of the Workshop on Computer Facilities for the Analysis of Protein and Nucleic Acid Sequence Information held March 1–3, 1979, Rockefeller University." (Appendix III of program for third workshop on need for nucleic acid sequence data bank, November 25, 1980.) [GenBank papers.]

Anderson, Herbert L. "Metropolis, Monte Carlo, and the MANIAC." *Los Alamos Science*, Fall 1986, 96–107.

Anderson, Philip W. "More Is Different." *Science* 177, no. 4047 (1972): 393–396.

Appadurai, Arjun. *The Social Life of Things: Commodities in Cultural Perspective*. Cambridge: Cambridge University Press, 1988.

Ashburner, Michael. "On the Representation of 'Gene Function' in Databases." Discussion paper for ISMB, Montreal, Version 1.2, June 19, 1998. ftp://ftp .geneontology.org/go/www/gene.ontology.discussion.shtml.

Bachmann, Charles W. "The Programmer as Navigator." *Communications of the ACM* 16, no. 11 (November 1973): 653–658. http://www.jdl.ac.cn/ turing/pdf/p653-bachman.pdf.

Bacon, Francis. "The New Atlantis." 1627. In *Francis Bacon, the Major Works*. Oxford: Oxford University Press, 2002.

———. *The New Organon*. 1620. Cambridge: Cambridge University Press, 2000.

Bada, Michael et al. "A Short Study on the Success of the Gene Ontology." *Web Semantics* 1, no. 2 (2004): 235–240.

Baldi, P., and S. Brunak. *Bioinformatics: The Machine Learning Approach*. Cambridge, MA: MIT Press, 1998.

Barabási, Albert-László, and Zoltán Oltvai. "Network Biology: Understanding the Cell's Functional Organization." *Nature Reviews Genetics* 5 (February 2004): 101–114.

Barnes, Barry, and John Dupré. *Genomes and What to Make of Them*. Chicago: University of Chicago Press, 2008.

Barton, Michael. "World of Bioinformatics Quest." *Bioinformatics Zen* [blog], January 28, 2008. http://www.bioinformaticszen.com/post/world-of -bioinformatics-quest:-character-generation/.

Baumlein, H. et al. "The Legumin Gene Family: Structure of a B Type Gene of *Vicia faba* and a Possible Legumin Gene Specific Regulatory Element." *Nucleic Acids Research* 14, no. 6 (1986): 2707–2720.

Beck, Ulrich. *Risk Society: Towards a New Modernity*. Thousand Oaks, CA: Sage, 1992.

Bell, George. Letter to George Cahill, April 22, 1986. [Box 7, Human Genome Project 1 (1986), WBG papers.]

Bentley, David R. et al. "Accurate Whole Human Genome Sequencing Using Reversible Terminator Chemistry." *Nature* 456 (2008): 53–59.

Berners-Lee, Tim. "Digital Future of the United States: Part I." Testimony before the Subcommittee on Telecommunications and the Internet of the House of Representatives Committee on Energy and Commerce, March 1, 2007.

———. "Transcript of Talk to Laboratory of Computer Science 35th Anniversary Celebrations." Massachusetts Institute of Technology, Cambridge, MA, April 14, 1999. http://www.w3.org/1999/04/13-tbl.html.

———. "The Web: Past, Present, and Future." August 1996. http://www.w3 .org/People/Berners-Lee/1996/ppf.html.

Berol, David. "Living Materials and the Structural Ideal: The Development of the Protein Crystallography Community in the 20th Century." PhD diss., Department of History, Princeton University, 2000.

Bertalanffy, Ludwig von. *General Systems Theory: Foundations, Development, Applications*. New York: G. Braziller, 1968.

Beyer, William A. et al. "A Molecular Sequence Metric and Evolutionary Trees." *Mathematical Biosciences* 19 (1974): 9–25.

Beynon, R. J. "CABIOS Editorial." *Computer Applications in the Biosciences: CABIOS* 1 (1985): 1.

Birney, Ewan. "An Interview with Ewan Birney" (keynote address at O'Reilly's Bioinformatics Technology Conference, October 18, 2001). http://www.oreillynet.com/pub/a/network/2001/10/18/birney.html.

Bishop, Martin J. "Software Club: Software for Molecular Biology. I. Databases and Search Programs." *BioEssays* 1, no. 1 (1984): 29–31.

Bishop, M., and C. Rawlings, eds. *DNA and Protein Sequence Analysis—A Practical Approach*. Oxford: IRL Press at Oxford University Press, 1997.

Black, Douglas L. "Mechanisms of Alternative Pre-messenger RNA Splicing." *Annual Review of Biochemistry* 72 (2003): 291–336.

Black, Grant C., and Paula E. Stephan. "Bioinformatics: Recent Trends in Programs, Placements, and Job Opportunities." Report to the Alfred P. Sloan Foundation, June 2004.

Blattner, Fred. "Second Workshop on Need for a Nucleic Acid Sequence Data Bank" (handwritten notes, October 27, 1980). [GenBank papers.]

Bolt, Beranek and Newman. "BBN Progress Report." (Excerpt from GenBank progress report no. 20, April to June 1985.) [GenBank papers.]

Boogerd, Fred C. et al. *Systems Biology: Philosophical Foundations*. Amsterdam: Elsevier, 2007.

Bork, Peer et al. "Predicting Function: From Genes to Genomes and Back." *Journal of Molecular Biology* 283 (1998): 707–725.

Bourdieu, Pierre. *Outline of a Theory of Practice*. Cambridge: Cambridge University Press, 1977.

Bowker, Geoffrey C., and Susan Leigh Star. *Sorting Things Out: Classification and Its Consequences*. Cambridge, MA: MIT Press, 1999.

Brenner, Sydney. "Sequence and Consequences." *Philosophical Transactions of the Royal Society B: Biological Sciences* 365, no. 1537 (2010): 207–212.

———. Address given at "GenBank Celebrates 25 Years of Service," Natcher Conference Center, National Institutes of Health, Bethesda, Maryland, April 8, 2008.

Broad Institute. "Philanthropists Eli and Edythe L. Broad Make Unprecedented Gift to Endow the Broad Institute of Harvard and MIT." Press release, September 4, 2008. http://www.broadinstitute.org/news/press-releases/1054.

———. "Rob Nicol." Accessed August 20, 2009. http://www.broadinstitute.org/about/bios/bio-nicol.html.

———. "Who Is Broad?" Accessed August 8, 2011. http://www.broadinstitute.org/what-broad/who-broad/who-broad.

Brown, P. O., and D. Botstein. "Exploring the New World of the Genome with DNA Microarrays." *Nature Genetics* 21 (1999): 33–37.

Burge, Christopher, and Sam Karlin. "Prediction of Complete Gene Structures

in Human Genomic DNA." *Journal of Molecular Biology* 268 (1997):
78–94.

Butler, Declan. "Are You Ready for the Revolution?" *Nature* 409 (February 15,
2001): 758–760.

Campbell-Kelly, Martin, and William Aspray. *Computer: A History of the
Information Machine.* New York: Basic Books, 1986.

Capel, Blanche et al. "Circular Transcripts of the Testis-Determining Gene *Sry*
in Adult Mouse Testis." *Cell* 73 (1993): 1019–1030.

Capshew, James H., and Karen A. Rader. "Big Science: Price to the Present."
Osiris 7 (1992): 3–25.

Caudevilla, Concha. "Natural Trans-Splicing in Carnitine Octanoyltransferase
Pre-mRNAs in Rat Liver." *Proceedings of the National Academy of Sciences
USA* 95, no. 21 (1998): 12185–12190.

Ceruzzi, Paul. *A History of Modern Computing.* Cambridge, MA: MIT Press,
1998.

———. *Reckoners: The Prehistory of the Digital Computer, from Relays to the
Stored Program Concept.* Westport, CT: Greenwood Press, 1983.

Chadarevian, Soraya de, and Nick Hopwood, eds. *Models: The Third Dimension of Science.* Stanford, CA: Stanford University Press, 2004.

Chang, Julia L. "Control and Optimization of *E. coli* Picking Process for DNA
Sequencing." Master's thesis, Sloan School of Management and Department
of Chemical Engineering, MIT, 2004.

Cinkosky, Michael J., and Jim Fickett. "A Relational Architecture for a Nucleotide Sequence Database" (December 18, 1986). [GenBank papers.]

Cinkosky, Michael J. et al. "A Technical Overview of the GenBank/HGIR
Database" (April 8, 1988). [GenBank papers.]

Cocquerelle, Claude et al. "Splicing with Inverted Order of Exons Occurs
Proximal to Large Introns." *EMBO Journal* 11, no. 3 (1992): 1095–1098.

Codd, Edgar F. "A Relational Model for Large Shared Data Banks." *Communications of the ACM* 13, no. 6 (June 1970): 377–387.

Collins, Harry M. "Does Electronic Communication Make Any Difference to
the Nature of Expertise." Paper presented at Science in the 21st Century:
Science, Society, and Information Technology, Waterloo, Ontario, September 8–12, 2008.

———. "The TEA Set: Tacit Knowledge and Scientific Networks." *Science
Studies* 4 (1974): 165–86.

Cook-Deegan, Robert M. "The Alta Summit, December 1984." *Genomics* 5
(1989): 661–663.

———. *The Gene Wars: Science, Politics, and the Human Genome.* New York:
W. W. Norton, 1994.

Cortada, James W. *Information Technology as Business History: Issues in the
History and Management of Computers.* Westport, CT: Greenwood Press,
1996.

Creager, Angela. *The Life of a Virus: Tobacco Mosaic Virus as an Experimental Model, 1930–1965.* Chicago: University of Chicago Press,
2002.

Dahl, Ole-Johan, and Kristen Nygaard. *Simula: A Language for Program-*

ming and Description of Discrete Events: Introduction and User's Manual, 5th ed. Oslo: Norwegian Computing Center, 1967.

Dandekar, B. et al. "Conservation of Gene Order: A Fingerprint of Physically Interacting Proteins." *Trends in Biotechnology Sciences* 23 (1998): 324–328.

Daston, Lorraine, and Peter Galison. "The Image of Objectivity." *Representations* 40 (Fall 1992): 81–128.

———. *Objectivity*. New York: Zone Books, 2007.

Davies, Kevin. "The DNA Data Deluge." *Bio-IT World* 7, no. 3 (April 2008): 22–28.

———. *The $1000 Genome: The Revolution in DNA Sequencing and the New Era of Personalized Medicine*. New York: Free Press, 2010.

Davis, Bernard D. "Sequencing the Human Genome: A Faded Goal." *Bulletin of the New York Academy of Medicine* 68, no. 1 (1992): 115–125.

Davison, Daniel B. et al. "Whither Computational Biology." *Journal of Computational Biology* 1, no. 1 (1994): 1–2.

Dayhoff, Margaret O. "Computer Aids to Protein Sequence Determination." *Journal of Theoretical Biology* 8 (1964): 97–112.

———. "Computer Analysis of Protein Evolution." *Scientific American* 221 (1969): 86–95.

———. "The Origin and Evolution of Protein Superfamilies." *Federation Proceedings* 35, no. 10 (August 1976): 2132–2138.

Dayhoff, Margaret O., and Richard V. Eck. *Atlas of Protein Sequence and Structure, 1967–68*. Silver Spring, MD: National Biomedical Research Foundation, 1968.

Dayhoff, Margaret O., and George E. Kimball. "Punched Card Calculation of Resonance Energies." *Journal of Chemical Physics* 17 (1949): 706–717.

Dayhoff, Margaret O., and Robert S. Ledley. "Comprotein: A Computer Program to Aid Primary Protein Structure Determination." *Proceedings of the American Federation of Information Processing Societies Joint Computer Conference*, Philadelphia, PA, December 4–6, 1962, 262–274.

Dayhoff, Margaret O. et al. "Evolution of Sequences within Protein Superfamilies." *Naturwissenschaften* 62 (1975): 154–161.

DeLisi, Charles. "Computers in Molecular Biology: Current Applications and Emerging Trends." *Science* 240, no. 4848 (1988): 47–52.

Department of Energy, Office of Health and Environmental Research. "Sequencing the Human Genome." (Summary report of the Santa Fe workshop, March 3–4 1986.) http://www.ornl.gov/sci/techresources/Human_Genome/publicat/1986santafereport.pdf.

Dietrich, Michael R. "Paradox and Persuasion: Negotiating the Place of Molecular Evolution within Evolutionary Biology." *Journal of the History of Biology* 31, no. 1 (1998): 85–111.

Doolittle, Russell F. et al. "Simian Sarcoma Virus *onc* Gene, *v-sis*, Is Derived from the Gene (or Genes) Encoding a Platelet-Derived Growth Factor." *Science* 221 (1983): 275–277.

———. "Some Reflections on the Early Days of Sequence Searching." *Journal of Molecular Medicine* 75 (1997): 239–241.

Dowell, Robin D. et al. "The Distributed Annotation System." *BMC Bioinformatics* 2, no. 7 (October 10, 2001).

Dubuisson, Olivier. *ASN.1: Communication between Heterogeneous Systems.* Trans. Philippe Fouquart (2000). Accessed March 30, 2009. http://www.oss.com/asn1/resources/books-whitepapers-pubs/asn1-books.html.

Durbin, Richard, and Jean Thierry-Mieg. "The ACEDB Genome Database" (1993). Accessed April 9, 2009. http://www.acedb.org/Cornell/dkfz.html.

Durbin, Richard et al. *Biological Sequence Analysis: Probabilistic Models of Proteins and Nucleic Acids.* Cambridge: Cambridge University Press, 1998.

Duster, Troy. "Race and Reification in Science." *Science* 307, no. 5712 (2005): 1050–1051.

Dyson, Freeman. "Our Biotech Future." *New York Review of Books* 54, no. 12 (July 19, 2007).

Edwards, Paul N. *The Closed World: Computers and the Politics of Discourse in the Cold War.* Cambridge, MA: MIT Press, 1996.

———. "Making History: New Directions in Computer Historiography." *IEEE Annals of the History of Computing* 23, no. 1 (2001): 85–87.

Eisen, M. B. et al. "Cluster Analysis and Display of Genome-Wide Expression Patterns." *Proceedings of the National Academy of Sciences USA* 95, no. 25 (1998): 14863–14868.

Eisenhower, Dwight D. "Farewell Radio and Television Address to the American People—17th January 1961." *Public Papers of the Presidents of the United States* (1960–1961): 1035–1040.

European Bioinformatics Institute. "Ensembl Release Coordination" (internal document, October 2008).

———. "The European Bioinformatics Institute (EBI)" (draft proposal, March 1992). ["EBI Background," GC papers.]

Ewing, B. et al. "Base Calling of Automated Sequence Traces Using Phred. I. Accuracy Assessment." *Genome Research* 8, no. 3 (1998): 175–185.

Feigenbaum, Lee et al. "The Semantic Web in Action." *Scientific American* 297 (December 2007): 90–97.

Fickett, James, and Christian Burks. "Development of a Database for Nucleotide Sequences" (draft, August 24, 1986). [GenBank papers.]

Field, Dawn et al. "Towards a Richer Description of Our Complete Collection of Genomes and Metagenomes: The 'Minimum Information about a Genome Sequence' (MIGS) Specification." *Nature Biotechnology* 26, no. 5 (2008): 41–47.

Fischer, Hans P. "Towards Quantitative Biology: Integration of Biological Information to Elucidate Disease Pathways and to Guide Drug Discovery." *Biotechnology Annual Review* 11 (2005): 1–68.

Forman, Paul. "Behind Quantum Electronics: National Security as a Basis for Physical Research in the United States, 1940–1960." *Historical Studies in the Physical and Biological Sciences* 18, no. 1 (1987): 149–229.

Fortun, Michael. "The Care of the Data." Accessed July 19, 2011. http://www.fortuns.org/?page_id=72.

———. "Projecting Speed Genomics." In *The Practices of Human Genetics,* edited by Michael Fortun and Everett Mendelsohn, 25–48. Boston, MA: Kluwer, 1999.

Fortun, Michael, and Sylvan S. Schweber. "Scientists and the Legacy of World War II: The Case of Operations Research (OR)." *Social Studies of Science* 23 (1993): 595–642.

Francoeur, Eric. "Cyrus Levinthal, the Kluge, and the Origins of Interactive Molecular Graphics." *Endeavour* 26, no. 4 (2002): 127–131.

———. "The Forgotten Tool: The Design and Use of Molecular Models." *Social Studies of Science* 27, no. 1 (1997): 7–40.

Francoeur, Eric, and Jérôme Segal. "From Model Kits to Interactive Computer Graphics." In *Models: The Third Dimension of Science*, edited by Nick Hopwood and Soraya de Chadarevian, 402–429. Palo Alto, CA: Stanford University Press, 2004.

Franklin, Laura R. "Exploratory Experiments." *Philosophy of Science* 72 (December 2005): 888–899.

Frantz, Simon A. et al. "Exon Repetition in mRNA." *Proceedings of the National Academy of Sciences USA* 96 (1999): 5400–5405.

Friedhoff, Richard M., and William Benzon. *Visualization: The Second Computer Revolution.* New York: Harry N. Abrams, 1989.

Friedland, Peter E. "Knowledge-Based Experimental Design in Molecular Genetics." PhD diss., Computer Science Department, School of Humanities and Sciences, Stanford University, 1979.

Friedland, Peter E. et al. "MOLGEN—Application of Symbolic Computing and Artificial Intelligence to Molecular Biology." n.d. [Hard binder, Box 63, EAF papers.]

Fry, Benjamin J. "Computational Information Design." PhD diss., Program in Media, Arts, and Sciences, School of Architecture and Planning, Massachusetts Institute of Technology, 2004.

Fujimura, Joan H., and Michael Fortun. "Constructing Knowledge across Social Worlds: The Case of DNA Sequence Databases in Molecular Biology." In *Naked Science: Anthropological Inquiry into Boundaries, Power, and Knowledge*, edited by Laura Nader, 160–173. New York: Routledge, 1996.

Galison, Peter. *How Experiments End.* Chicago: University of Chicago Press, 1987.

———. *Image and Logic: A Material Culture of Microphysics.* Chicago: University of Press, 1997.

Galperin, Michael Y., and Guy R. Cochrane. "Nucleic Acids Research Annual Database Issue and the NAR Online Molecular Biology Database Collection." *Nucleic Acids Research* 37, database issue (2009): D1–D4.

Garcia-Sancho, Miguel. "From the Genetic to the Computer Program: The Historicity of 'Data' and 'Computation' in the Investigations of the Nematode Worm C. elegans (1963–1998)." *Studies in the History and Philosophy of the Biological and Biomedical Sciences* 43 (2012): 16–28.

Gaudillière, Jean-Paul, and Hans-Jörg Rheinberger, eds. *Classical Genetic Research and Its Legacy: The Mapping Cultures of Twentieth-Century Genetics.* New York: Routledge, 2004.

Gene Ontology Consortium. "Gene Ontology: Tool for the Unification of Biology." *Nature Genetics* 25 (May 2000): 25–29.

Gibbs, Adrian J., and George A. McIntyre. "The Diagram, a Method for Comparing Sequences: Its Use with Amino Acid and Nucleotide Sequences." *European Journal of Biochemistry* 16, no. 1 (1970): 1–11.

Gieryn, Thomas F. "Two Faces on Science: Building Identities for Molecular Biology and Biotechnology." In *The Architecture of Science*, edited by Peter Galison and Emily Thompson (Cambridge, MA: MIT Press, 1999): 424–455.

Gilbert, Jack A. et al. "Meeting Report: The Terabase Metagenomics Workshop and the Vision of an Earth Microbiome Project." *Standards in Genomic Science* 3, no. 3 (2010): 243–248.

Gilbert, Walter. "Towards a Paradigm Shift in Biology." *Nature* 349 (January 10, 1991): 99.

Gillies, D. A. "Popper and Computer Induction." *BioEssays* 23, no. 9 (2001): 859–860.

Gingeras, Thomas R., and Richard J. Roberts. "Steps toward Computer Analysis of Nucleotide Sequences." *Science* 209 (September 19, 1980): 1322–1328.

Giustini, Dean. "Web 3.0 and Medicine." *British Medical Journal* 35 (2007): 1273–1274.

Glaser, Donald. "The Bubble Chamber, Bioengineering, Business Consulting, and Neurobiology." Interview by Eric Vettel, Regional Oral History Office, Bancroft Library, University of California, Berkeley, 2006.

Glass, John I. et al. "Essential Genes of a Minimal Bacterium." *Proceedings of the National Academy of Sciences USA* 103, no. 2 (2006): 425–430.

Goad, Walter B. "Sequence Analysis: Contributions by Ulam to Molecular Genetics." In *From Cardinals to Chaos: Reflections on the Life and Legacy of Stanislaw Ulam*, edited by Necia Grant Cooper et al., 288–293. Cambridge: Cambridge University Press, 1989.

———. "T-division" (office memorandum to T-division leader search committee, November 29, 1972, Box 1, Los Alamos Reports 1972). [WBG papers.]

———. "A Theoretical Study of Extensive Cosmic Ray Air Showers." PhD diss., Duke University, Department of Physics, 1953. [Box 2, WBG papers.]

———. Undated letter to Japan[?]. [Box 12, Laboratory notes, n.d., WBG papers.]

———. "Vita" (January 30, 1974, Box 1, Los Alamos Reports 1974). [WBG papers.]

———. "Wat: A Numerical Method for Two-Dimensional Unsteady Fluid Flow." LASL Report, LAMS-2365 (1960). [Box 2, Publications 1 (1952–1965), WBG papers.]

Goad, Walter B., and John R. Cann. "Theory of Moving Boundary Electrophoresis of Reversibly Interacting Systems." *Journal of Biological Chemistry* 240, no. 1 (1965): 148–155.

———. "Theory of Zone Electrophoresis of Reversibly Interacting Systems." *Journal of Biological Chemistry* 240, no. 3 (1965): 1162–1164.

———. "The Theory of Transport of Interacting Systems of Biological Macromolecules." *Advances in Enzymology* 30 (1968): 139–175.

Goad, Walter B., and R. Johnson. "A Montecarlo Method for Criticality Problems." *Nuclear Science and Engineering* 5 (1959): 371–375.

Goetz, Thomas. "Craig Venter Wants to Solve the World's Energy Crisis." *Wired* 20, no. 6 (2012): 108–114.

———. "23andMe Will Decode Your Genome for $1000. Welcome to the Age of Genomics." *Wired* 15, no. 12 (2007).

Goffman, Erving. *The Presentation of Self in Everyday Life*. New York: Doubleday, 1959.

Gusfield, Dan, *Algorithms on Strings, Trees, and Sequences: Computer Science and Computational Biology*. Cambridge: Cambridge University Press, 1997.

Hagen, Joel B. "The Introduction of Computers into Systematic Research in the United States during the 1960s." *Studies in the History and Philosophy of Biology and Biomedical Sciences* 32, no. 2 (2001): 291–314.

———. "Naturalists, Molecular Biologists, and the Challenge of Molecular Evolution." *Journal of the History of Biology* 32, no. 2 (1999): 321–341.

Haigh, Thomas. "The Chromium-Plated Tabulator: Institutionalizing an Electronic Revolution, 1954–1958." *IEEE Annals of the History of Computing* 23, no. 4 (2001): 75–104.

———. "'A Veritable Bucket of Facts': Origins of the Data Base Management System." *SIGMOD Record* 35, no. 2 (June 2006): 33–49.

Han, Jing Dong J. et al. "Effect of Sampling on Topology Predictions of Protein-Protein Interaction Networks." *Nature Biotechnology* 23 (2005): 839–844.

Harris, Timothy D. et al. "Single-Molecule DNA Sequencing of a Viral Genome." *Science* 320, no. 5872 (2008): 106–109.

Harte, Rachel, and Mark Diekhans. "Fishing for Genes in the UCSC Browser: A Tutorial" (University of Santa Cruz, September 18, 2008). Accessed April 22, 2009. http://genome.ucsc.edu/training.html.

Helmreich, Stefan. *Silicon Second Nature: Culturing Artificial Life in a Digital World*. Berkeley: University of California Press, 1998.

Higginbotham, Julie S. "Biomedical Facility Shows Best of Modern Lab Design." *R&D*, June 1, 2011. http://www.rdmag.com/Lab-Design-News/Articles/2011/06/Lab-Of-The-Year-Biomedical-Facility-Shows-Best-Of-Modern-Lab-Design/.

———. "Collaborative Venture Produces Versatile Open Laboratory." *R&D*, May 24, 2007. http://www.rdmag.com/Awards/Lab-Of-The-Year/2007/05/Collaborative-Venture-Produces-Versatile-Open-Laboratory/.

Hilgartner, Stephen. "Biomolecular Databases: New Communications Regimes for Biology?" *Science Communication* 17, no. 2 (1995): 240–263.

———. "Data Access Policy in Genome Research." In *Private Science: Biotechnology and the Rise of the Molecular Sciences*, edited by Arnold Thackray, 202–218. Philadelphia: University of Pennsylvania Press, 1995.

Himmelreich, R. et al. "Comparative Analysis of the Genomes *Mycoplasma pneumoniae* and *Mycoplasma genitalium.*" *Nucleic Acids Research* 25 (1997): 701–712.

Hodges, Andrew. *Alan Turing: The Enigma.* New York: Simon and Schuster, 1983.

Hogeweg, Paulien. "Simulating the Growth of Cellular Forms." *Simulation* 31 (1978): 90–96.

Honeybee Genome Sequencing Consortium. "Insights into Social Insects from the Genome of the Honeybee *Apis mellifera.*" *Nature* 443 (2006): 931–949.

Hopkin, Karen. "Bring Me Your Genomes." *Scientist* 19, no. 11 (2005): 60–61.

Hubbard, Tim et al. "The Ensembl Genome Database Project." *Nucleic Acids Research* 30, no. 1 (2002): 38–41.

Human Microbiome Project Consortium. "A Framework for Human Microbiome Research." *Nature* 486, no. 7402 (2012): 215–221.

———. "Structure, Function, and Diversity of the Healthy Human Microbiome." *Nature* 486, no. 7402 (2012): 207–214.

Hunt, Lois. "Margaret Oakley Dayhoff, 1925–1983." *Bulletin of Mathematical Biology* 46, no. 4 (1984): 467–472.

Hunter, Lawrence E. "Artificial Intelligence and Molecular Biology." *AI Magazine* 11, no. 5 (1993): 27–36.

Huynen, M. A. et al. "Genomics: Differential Genome Analysis Applied to the Species Specific Features of *Helicobacter pylori.*" *FEBS Letters* 426 (1998): 1–5.

International Society for Computational Biology. "History of ISCB." Accessed June 9, 2009. http://www.iscb.org/iscb-history.

Irish, J. M. et al. "Single Cell Profiling of Potentiated Phospho-Protien Networks in Cancer Cells." *Cell* 118 (July 23, 2004): 217–228.

Jordan, Elke. Letter to Director, Office of Human Genome Research, March 26, 1989. [GenBank papers.]

———. "Record of the Meeting between Drs. Watson and Lindberg on 11/9/88" (memorandum to record, November 15, 1988). [GenBank papers.]

———. "Request for Contract Action" (memorandum to Steven Thornton, Executive Officer, NIGMS, July 10 1981). ["GenBank," JAL papers.]

Kabat, Elvin. "Minutes from GenBank Advisory Meeting" (October 31, 1985). ["Genbank. Advisory Group. Minutes and Notes 2 (1985)," WBG papers.]

Kaiser, David. "The Postwar Suburbanization of American Physics." *American Quarterly* 56 (December 2004): 851–888.

———. "Scientific Manpower, Cold War Requisitions, and the Production of American Physicists after World War II." *Historical Studies in the Physical and Biological Sciences* 33 (Fall 2002): 131–159.

Kaufmann, Stuart. *The Origins of Order: Self-Organization and Selection in Evolution.* Oxford: Oxford University Press, 1993.

Kay, Lily E., *Who Wrote the Book of Life? A History of the Genetic Code.* Stanford, CA: Stanford University Press, 2000.

Kell, Douglas B., and Stephen G. Oliver. "Here Is the Evidence, Now What Is the Hypothesis? The Complementary Roles of Inductive and Hypothesis-

Driven Science in the Post-genomic Era." *BioEssays* 26, no. 1 (2005): 99–105.

Kelleher, Kevin. "Google, Sergey, and 23andMe: Why It All Makes Sense." *Gigaom* (May 24, 2007). http://gigaom.com/2007/05/24/google-sergey-and-23andme-why-it-all-makes-sense/.

Keller, Evelyn Fox. *Century of the Gene*. Cambridge, MA: Harvard University Press, 2000.

———. *Making Sense of Life: Explaining Biological Development with Models, Metaphors, and Machines*. Cambridge, MA: Harvard University Press, 2002.

Kent, W. James et al. "The Human Genome Browser at UCSC." *Genome Research* 12 (2002): 996–1006.

Kent, W. James, and Heidi Brumbaugh. "autoSql and autoXml: Code Generators for the Genome Project." *Linux Journal* (July 1, 2002). Accessed April 21, 2009. http://www.linuxjournal.com/article/5949.

Kevles, Daniel J. "Big Science and Big Politics in the United States: Reflections on the Death of the SSC and the Left of the Human Genome Project." *Historical Studies in the Physical and Biological Sciences* 27 (1997): 269–297.

Kirby, Maurice W. *Operational Research in War and Peace*. London: Imperial College Press, 2003.

Kirschstein, Ruth, and James C. Cassatt. "Source Selection" (memorandum to contracting officer RFP NIH-GM-87-04, August 27, 1987). [GenBank papers.]

Knight, Jonathan. "Bridging the Culture Gap." *Nature* 419 (September 19, 2002): 244–246.

Koerner, Lisbet. *Linnaeus: Nature and Nation*. Cambridge, MA: Harvard University Press, 1999.

Kohler, Robert. *Lords of the Fly: Drosophila Genetics and the Experimental Life*. Chicago: University of Chicago Press, 1994.

Koonin, E. V. et al. "Comparison of Archaeal and Bacterial Genomes: Computer Analysis of Protein Sequences Predicts Novel Functions and Suggests a Chimeric Origin for the Archaea." *Molecular Microbiology* 25 (1997): 619–637.

Landecker, Hannah. "Microcinematography and the History of Science and Film." *Isis* 97, no. 1 (2006): 121–132.

Latour, Bruno, and Steve Woolgar. *Laboratory Life: The Construction of Scientific Facts*. Beverly Hills: Sage, 1979.

Ledford, H. "Molecular Biology gets Wikified." *Nature Online* (July 23, 2008). doi:10.1038/News.2008.971.

Ledley, Robert S. "Digital Electronic Computers in Biomedical Science." *Science* 130 (1959): 1232.

———. *Use of Computers in Biology and Medicine*. New York: McGraw-Hill, 1965.

Lenoir, Timothy. "Science and the Academy of the 21st Century: Does Their Past Have a Future in the Age of Computer-Mediated Networks?" Paper presented at Ideale Akademie: Vergangene Zukunft oder konkrete Utopie?, Berlin Akademie der Wissenschaften, May 12, 2000.

————. "Shaping Biomedicine as an Information Science." In *Proceedings of the 1998 Conference on the History and Heritage of Science Information Systems*, edited by Mary Ellen Bowden et al., 27–45. Medford, NJ: Information Today, 1999.

Leonelli, Sabina. "Centralizing Labels to Distribute Data: The Regulatory Role of Genomic Consortia." In *Handbook of Genetics and Society: Mapping the New Genomic Era*, edited by Paul Atkinson et al. London: Routledge, 2009.

————. "Introduction: Making Sense of Data-Driven Research in the Biological and Biomedical Sciences." *Studies in the History and Philosophy of the Biological and Biomedical Sciences* 43 (2012): 1–3.

Lesk, Arthur M., ed. *Computational Molecular Biology: Sources and Methods for Sequence Analysis*. Oxford: Oxford University Press, 1988.

Levinthal, Cyrus et al. "Computer Graphics in Macromolecular Chemistry." In *Emerging Concepts in Computer Graphics*, edited by Don Secrest and Jurg Nievergelt, 231–253. New York: W. A. Benjamin, 1968.

Levy, S. et al. "The Diploid Genome Sequence of an Individual Human." *PLoS Biology* 5, no. 10 (2007): e254.

Lewin, Roger. "DNA Databases Are Swamped." *Science* 232 (June 27, 1986): 1599.

Lewis, B. P. et al. "Prediction of Mammalian MicroRNA Targets." *Cell* 115, no. 7 (2003): 787–798.

Lewis, Suzanna E. "Gene Ontology: Looking Backwards and Forwards." *Genome Biology* 6 (2005): 103–106.

Lohr, Steve. "Microsoft Rolls Out Personal Health Records." *New York Times*, October 4, 2007.

Los Alamos National Laboratory. "Database Working Group Meeting with David Lipman" (March 24, 1989). [GenBank papers.]

————. "Los Alamos Progress Report" (January to March 1985). [GenBank papers.]

————. "T-10 Theoretical Biology and Biophysics" (1977). [Box 1, Los Alamos reports 1977, WBG papers.]

————. "$2 Million Earmarked for Genetic Data Bank at LANL." (*LANL News Bulletin*, 1982.) [Box 11, Articles about Goad—from Los Alamos Bulletin (1982–1987), WBG papers.]

Loua, Toussaint. *Atlas statistique de la population de Paris*. Paris, J. Dejey, 1873.

Mackenzie, Adrian. "Bringing Sequences to Life: How Bioinformatics Corporealizes Sequence Data." *New Genetics and Society* 22, no. 3 (2003): 315–332.

Mackenzie, Donald. "The Influence of the Los Alamos and Livermore National Laboratories on the Development of Supercomputing." *Annals of the History of Computing* 13 (1991): 179–201.

Mardis, Elaine. "The Impact of Next-Generation Sequencing Technology on Genetics." *Trends in Genetics* 24, no. 3 (March 2008): 133–141.

Maruyama, Kazunori. "Genome Sequencing Technology: Improvement of Electrophoretic Sequencing Process and Analysis of the Sequencing Tool

Industry." Master's thesis, Sloan School of Management and Department of Chemical Engineering, MIT, 2005.

Matlin, Arianne J. et al. "Understanding Alternative Splicing: Towards a Cellular Code." *Nature Reviews Molecular and Cellular Biology* 6 (May 2005): 386–398.

Maturana, Humberto, and Francisco Varela. "Autopoiesis: The Organization of the Living." In *Autopoiesis and Cognition: The Realization of the Living.* Boston, MA: D. Reidel, 1980.

McCormick, Bruce H. et al., eds. "Visualization in Scientific Computing." *Computer Graphics* 21, no. 6 (November 1987).

Metropolis, Nicholas. "The Los Alamos Experience, 1943–45." In *A History of Scientific Computing*, edited by Stephen G. Nash, 237–250. New York: ACM Press, 1990.

Metropolis, Nicholas, and C. Nelson. "Early Computing at Los Alamos." *Annals of the History of Computing* 4, no. 4 (1982): 348–357.

Miller, Daniel. *Material Culture and Mass Consumption.* New York: Wiley-Blackwell, 1987.

Morgan, Mary S., and Peter Howlett, eds. *How Well Do Facts Travel? The Dissemination of Reliable Knowledge.* Cambridge: Cambridge University Press, 2010.

Mount, David W. *Bioinformatics: Sequence and Genome Analysis.* 2nd ed. Cold Spring Harbor, NY: Cold Spring Harbor Laboratory Press, 2004.

Mushegian, A. R., and E. V. Koonin. "Gene Order Is Not Conserved in Bacterial Evolution." *Trends in Genetics* 12 (1996): 289–290.

———. "A Minimal Gene Set for Cellular Life Derived by Comparison of Complete Bacterial Genomes." *Proceedings of the National Academy of Sciences USA* 93 (1996): 10268–10273.

Myers, Natasha. "Modeling Proteins, Making Scientists: An Ethnography of Pedagogy and Visual Cultures in Contemporary Structural Biology." PhD diss., HASTS, Massachusetts Institute of Technology, 2007.

National Academy of Sciences, National Academy of Engineering, Institute of Medicine, Committee on Science, Engineering, and Public Policy. *Information Technology and the Conduct of Research: The User's View.* Report of the Panel on Information Technology and the Conduct of Research. Washington, DC: National Academy Press, 1989.

National Biomedical Research Foundation. "Rebuttal to the Argument against the Switch" (undated, circa 1982). [GenBank papers.]

National Center for Human Genome Research. "Options for Management of Informatics Related to the Human Genome Project" (undated). [GenBank papers.]

National Human Genome Research Institute. "Genome Sequencing Centers (U54)" (Request for Applications for NHGRI grants, December 22, 2005). http://grants.nih.gov/grants/guide/rfa-files/RFA-HG-06-001.html.

National Institute of General Medical Science and National Library of Medicine. "First Status Report Regarding GenBank" (memorandum to Acting Director, NIH, December 1, 1989). [GenBank papers.]

National Institutes of Health. "Agenda: 3rd Workshop on the Need for a Nucleic Acid Sequence Data Base" (December 1980). [GenBank papers.]

———. "Funding Plan: Nucleic Acid Sequence Database" (June 12, 1981). ["GenBank," JAL papers]

———. "MOLGEN Report" (to Elke Jordan, September 8, 1980). [EAF papers.]

———. "Proposed Statement of Work" (undated, circa 1987). [GenBank papers.]

———. "Report and Recommendation of the Nucleic Acid Sequence Database Workshop" (Bethesda, MD, July 14–15, 1980). ["Data Bank History," GC papers.]

National Institutes of Health, National Library of Medicine. "Long Range Plan." (Report of the Board of Regents, January 1987.)

———. "Talking One Genetic Language: The Need for a National Biotechnology Information Center" (Bethesda, MD, 1987).

Nature. "Community Cleverness Required." Editorial. *Nature* 455, no. 7209 (2008): 1.

———. "An Eye for Success." Editorial. *Nature* 478, no. 7368 (2011): 155–156.

———. "Post-genomic Cultures." Editorial. *Nature* 409, no. 545 (February 1, 2001): 545.

Needleman, Saul B., and Christian D. Wunsch. "A General Method Applicable to the Search for Similarities in the Amino-Acid Sequence of Two Proteins." *Journal of Molecular Biology* 48, no. 3 (1970): 443–453.

Nelkin, Dorothy, and Susan Lindee. *The DNA Mystique: The Gene as a Cultural Icon.* New York: W. H. Freeman, 1996.

Neumann, Eric K. "Semantic Web." Accessed September 1, 2008. http://eneumann.org/.

Ng, Pauline C. et al. "An Agenda for Personalized Medicine." *Nature* 461 (October 8, 2009): 724–726.

Nigro, Janice M. et al. "Scrambled Exons." *Cell* 64 (1991): 607–613.

Novas, Carlos, and Nikolas Rose. "Genetic Risk and the Birth of the Somatic Individual." *Economy and Society* 29, no. 4 (2000): 485–513.

November, Joseph A. *Biomedical Computing: Digitizing Life in the United States.* Baltimore, MD: Johns Hopkins University Press, 2012.

———. "Digitizing Life: The Introduction of Computers to Biology and Medicine." PhD diss., Department of History, Princeton University, 2006.

OBO Foundry. "The Open Biological and Biomedical Ontologies." Updated August 11, 2011. http://www.obofoundry.org/.

Ohtsubo, Eiichi, and Daniel B. Davison. Letter to Elke Jordan, November 25, 1980. [GenBank papers.]

Olsen, Maynard V. "A Time to Sequence." *Science* 270, no. 5235 (1995): 344–396.

The 1000 Genomes Project Consortium. "An Integrated Map of Genetic Variation from 1092 Human Genomes." *Nature* 491 (November 1, 2012): 56–65.

O'Reilly, Tim et al. "The Importance of Perl." O'Reilly Media, 1998. http:// oreillynet.com/pub/a/oreilly/perl/news/importance_0498.html.

Ostell, James M. "Evolution of a Molecular Biology Software Package with Applications." PhD diss., Department of Cellular and Developmental Biology, Harvard University, 1987.

———. "Integrated Access to Heterogeneous Data from NCBI." *IEEE Engineering in Medicine and Biology* (November/December 1995): 730–736.

———. "A Hobby Becomes a Life." Computational Biology at McMaster University, Ontario, Canada. Accessed June 4, 2009. http://life.biology .mcmaster.ca/~brian/biomath/role.Ostell.html.

Ostell, James M. et al. "The NCBI Data Model." In *Bioinformatics: A Practical Guide to the Analysis of Genes and Proteins*, 2nd edition, edited by Andreas D. Baxevanis and B. F. Francis Ouellette, 19–43. New York: John Wiley and Sons, 2001.

Parry, Bronwyn, *Trading the Genome: Investigating the Commodification of Bio-Information*. New York: Columbia University Press, 2004.

Pauling, Linus, and Emile Zuckerkandl. "Divergence and Convergence in Proteins." In *Evolving Genes and Proteins*, edited by Vernon Bryson and Henry J. Vogel, 97–166. San Diego, CA: Academic Press, 1965.

Person, Kerry P. "Operational Streamlining in a High-Throughput Genome Sequencing Center." Master's thesis, Sloan School of Management and Department of Chemical Engineering, MIT, 2006.

Popper, Karl. *The Logic of Scientific Discovery*. London: Routledge, [1935] 1999.

Pottage, Alain. "Too Much Ownership: Bio-prospecting in the Age of Synthetic Biology." *Biosocieties* 1 (2006): 137–158.

Price, Derek J. de Solla. *Little Science, Big Science . . . and Beyond*. Oxford: Oxford University Press, 1963.

Prusinkiewicz, P., and A. Lindenmeyer. *The Algorithmic Beauty of Plants*. Berlin: Springer-Verlag, 1990.

Pustell, James, and Fotis C. Kafatos. "A Convenient and Adaptable Microcomputer Environment for DNA and Protein sequence Manipulation and Analysis." *Nucleic Acids Research* 14, no. 1 (1986): 479–488.

———. "A Convenient and Adaptable Package of Computer Programs for DNA and Protein Sequence Management, Analysis and Homology Determination." *Nucleic Acids Research* 12, no. 1 (1984): 643–655.

———. "A Convenient and Adaptable Package of DNA Sequence Analysis Programs for Microcomputers." *Nucleic Acids Research* 10, no. 1 (1982): 51–58.

Rabinow, Paul. "Artificiality and Enlightenment: From Sociobiology to Biosociality." In *Zone 6: Incorporations*, edited by J. Crary, 234–252. Cambridge, MA: MIT Press, 1992.

Rader, Karen. *Making Mice: Standardizing Animals for American Biomedical Research, 1900–1955*. Princeton, NJ: Princeton University Press, 2004.

Rajan, Kaushik S. *Biocapital: The Constitution of Postgenomic Life*. Durham, NC: Duke University Press, 2006.

Rajewsky, N. "MicroRNA Target Prediction in Animals." *Nature Genetics* 38, supplement (2006): S8–S13.

Rasmussen, Nicolas. *Picture Control: The Electron Microscope and the Transformation of Biology, 1940–1960*. Stanford, CA: Stanford University Press, 1997.

Reas, Casey, and Ben Fry. *Processing: A Programming Handbook for Visual Designers and Artists*. Cambridge, MA: MIT Press, 2007.

Rheinberger, Hans-Jörg, and Jean-Paul Gaudillière, eds. *From Molecular Genetics to Genomics: The Mapping Cultures of Twentieth Century Genetics*. New York: Routledge, 2004.

Rigatti, Roberto et al. "Exon Repetition: A Major Pathway for Processing mRNA of Some Genes is Allele Specific." *Nucleic Acids Research* 32, no. 2 (2004): 441–446.

Rioux, John D. et al. "Genome-Wide Association Study Identifies New Susceptibility Loci for Crohn Disease and Implicates Autophagy in Disease Pathogenesis." *Nature Genetics* 39 (2007): 596–604.

Rommens, J. M. et al. "Identification of the Cystic Fibrosis Gene: Chromosome Walking and Jumping." *Science* 245 (1989): 1059–1065.

Roos, David S. "Bioinformatics—Trying to Swim in a Sea of Data." *Science* 291 (2001): 1260–1261.

Rose, Nicolas. *The Politics of Life Itself: Biomedicine, Power, and Subjectivity in the Twenty-First Century*. Princeton, NJ: Princeton University Press, 2007.

Rosenberg, Charles E., and Janet Golden. *Framing Disease: Studies in Cultural History*. New Brunswick, NJ: Rutgers University Press, 1992.

Rosenberg, Scott A. "Managing a Data Analysis Production Line: An Example from the Whitehead/MIT Center for Genomic Research." Master's thesis, Sloan School of Management and Department of Electrical Engineering/ Computer Science, MIT, 2003.

Rothberg, Jonathan M., and John H. Leamon. "The Development and Impact of 454 Sequencing." *Nature Biotechnology* 26, no. 10 (October 2008): 1117–1124.

Rothberg, Jonathan M. et al. "An Integrated Semiconductor Device Enabling Non-optical Genome Sequencing." *Nature* 475 (July 21, 2011): 348–352.

Sandberg, Rickard et al. "Proliferating Cells Express mRNAs with Shortened 3′ Untranslated Regions and Fewer MicroRNA Target Sites." *Science* 320, no. 5883 (2008): 1643–1647.

Sanger Centre. "The Farm FAQ—Sanger Wiki." Accessed November 3, 2008. http://scratchy.internal.sanger.ac.uk/wiki/index.php/The_Farm_FAQ.

Schlereth, Thomas J., ed. *Material Culture in America*. Walnut Creek, CA: Altamira Press, 1999.

Schneider, Tom. "Suggestions for GenBank" (October 31, 1985). ["Genbank. Advisory Group. Minutes and Notes 2 (1985)," WBG papers.]

Schulze-Kremer, S. "Ontologies for Molecular Biology." *Pacific Symposium on Biocomputing '98* (1998): 695–706.

Scott, James C. *Seeing Like a State: How Certain Schemes to Improve the Human Condition Have Failed*. New Haven, CT: Yale University Press, 1998.

Setubal, João C., and João Medianas. *Introduction to Computational Molecular Biology*. Boston, MA: PWS Publishing, 1997.

Shao, X. et al. "Bioinformatic Analysis of Exon Repetition, Exon Scrambling and Trans-splicing in Humans." *Bioinformatics* 22, no. 6 (2006): 692–698.

Shapin, Steven. "The House of Experiment in Seventeenth-Century England." *Isis* 79 (3, Special Issue on Artifact and Experiment, September 1988): 373–404.

———. "The Invisible Technician." *American Scientist* 77 (1989): 554–563.

———. "Pump and Circumstance: Robert Boyle's Literary Technology." *Social Studies of Science* 14 (1984): 481–520.

Sherman, B. T. et al. "DAVID Knowledgebase: A Gene-Centered Database Integrating Heterogeneous Annotation Resources to Facilitate High-Throughput Gene Functional Analysis." *BMC Bioinformatics* 8 (November 2, 2007): 426.

Silverberg, Michael. "The Glass Lab." *Metropolismag.com*, February 14, 2007. http://www.metropolismag.com/cda/story.php?artid=2517.

Smalheiser, N. R. "Informatics and Hypothesis-Driven Research." *EMBO Reports* 3 (2002): 702.

Smith, Barry. "Ontology." In *Blackwell Guide to the Philosophy of Computing and Information*, edited by L. Floridi, 155–166. Malden, MA: Blackwell, 2003.

———. "Ontology as the Core Discipline of Biomedical Informatics: Legacies of the Past and Recommendations for the Future Direction of Research." In *Computing, Philosophy, and Cognitive Science*, edited by G. D. Crnkovic and S. Stuart. Cambridge: Cambridge Scholars Press, 2006.

———. "What Is an Ontology and What Is It Useful For?" (videorecorded lecture). Accessed February 13, 2009. http://ontology.buffalo.edu/smith/Ontology_Course.html.

Smith, Brian Cantwell. *On the Origin of Objects*. Cambridge, MA: MIT Press, 1996.

Smith, Kent A. "Laws, Leaders, and Legends of the Modern National Library of Medicine." *Journal of the Medical Library Association* 96, no. 2 (2008): 121–133.

Smith, Michael K. et al. "OWL Web Ontology Language Guide" (W3C recommendation, February 10, 2004). http://www.w3.org/TR/owl-guide/.

Smith, Temple F. "The History of the Genetic Sequence Databases." *Genomics* 6 (1990): 701–707.

Smith, Temple F., and Michael S. Waterman. "Identification of Common Molecular Subsequences." *Journal of Molecular Biology* 147 (1981): 195–197.

Sneath, Peter H. A. "The Application of Computers to Taxonomy." *Journal of General Microbiology* 17 (1957): 201–226.

Stabenau, Arne et al. "The Ensembl Core Software Libraries." *Genome Research* 14 (2004): 929–933.

Stacy, Ralph W., and Bruce D. Waxman, eds. *Computers in Biomedical Research*. Vol. 1. New York: Academic Press, 1965.

Stalker, James et al. "The Ensembl Web Site: Mechanics of a Genome Browser." *Genome Research* 14 (2004): 951–955.

Stamatis, D. H. *Six Sigma Fundamentals: A Complete Guide to the Systems, Methods, and Tools*. New York: Productivity Press, 2004.

Star, Susan L., and James R. Griesemer. "Institutional Ecology, 'Translations,' and Boundary Objects: Amateurs and Professionals in Berkeley's Museum of Vertebrate Zoology, 1907–1939." *Social Studies of Science* 19, no. 4 (1989): 387–420.

Stark, Alexander et al. "Discovery of Functional Elements in 12 *Drosophila* Genomes Using Evolutionary Signatures." *Nature* 450 (November 8, 2007): 219–232.

Stein, Lincoln. "The Case for Cloud Computing in Genome Informatics." *Genome Biology* 11, no. 5 (2010): 207.

———. "How Perl Saved the Human Genome Project." *Perl Journal* (September 1996). http://www.bioperl.org/wiki/How_Perl_saved_human_genome.

Stephan, Paula E., and Grant C. Black. "Hiring Patterns Experienced by Students Enrolled in Bioinformatics/Computational Biology Programs." Report to the Alfred P. Sloan Foundation, 1999.

Stevens, Hallam. "Coding Sequences: A History of Sequence Comparison Algorithms as a Scientific Instrument." *Perspectives on Science* 19, no. 3 (2011): 263–299.

———. "On the Means of Bio-production: Bioinformatics and How to Make Knowledge in a High-Throughput Genomics Laboratory." *Biosocieties* 6 (2011): 217–242.

Stewart, Bruce. "An Interview with Lincoln Stein." *O'Reilly* (December 7, 2001). http://www.oreillynet.com/pub/a/network/2001/12/07/stein.html.

Strasser, Bruno J. "Collecting and Experimenting: The Moral Economies of Biological Research, 1960s–1980s." Workshop at Max-Planck-Institut fur Wissenschaftsgeschichte, History and Epistemology of Molecular Biology and Beyond: Problems and Perspectives, October 13–15, 2005: 105–123.

———. "Collecting, Comparing, and Computing Sequences: The Making of Margaret O. Dayhoff's *Atlas of Protein Sequence and Structure*." *Journal of the History of Biology* 43, no. 4 (2010): 623–660.

———. "Data-Driven Sciences: From Wonder Cabinets to Electronic Databases." *Studies in the History and Philosophy of Biological and Biomedical Sciences* 43 (2012): 85–87.

———. "The Experimenter's Museum: GenBank, Natural History, and the Moral Economies of Biomedicine." *Isis* 102, no. 1 (2011): 60–96.

———. "GenBank: Natural History in the 21st Century?" *Science* 322 (2008): 537–538.

Strathern, Marilyn, *After Nature: English Kinship in the Late Twentieth Century*. Cambridge: Cambridge University Press, 1992.

Suárez-Díaz, Edna, and Victor H. Anaya-Muñoz. "History, Objectivity, and the Construction of Molecular Phylogenies." *Studies in the History and Philosophy of Science Part C* 39, no. 4 (2008): 451–468.

Sulston, John et al. "Software for Genome Mapping by Fingerprinting Techniques." *Computer Applications in the Biosciences* 4, no. 1 (1988): 125–132.

Taylor, Crispin. "Bioinformatics: Jobs Galore." *Science Careers*, Septem-

ber, 1, 2000. http://sciencecareers.sciencemag.org/career_magazine/
previous_issues/articles/2000_09_01/noDOI.8547776132534311194.

Thacker, Eugene, *Biomedia*. Minneapolis: University of Minnesota Press, 2004.

Theoretical Biology and Biophysics Group. "Proposal to Establish a National Center for Collection, and Computer Storage and Analysis of Nucleic Acid Sequences." (University of California, LASL, December 17, 1979.) [Box 2, Publications 3 (1978–1981), WBG papers.]

Thompson, D'Arcy. *On Growth and Form*. New York: Dover, 1917.

Tomlinson, John. *Globalization and Culture*. Chicago: University of Chicago Press, 1999.

Turkle, Sherry, ed. *Evocative Objects: Things We Think With*. Cambridge, MA: MIT Press, 2011.

Turner, Stephen. "Video Tip: Use Ensembl BioMart to Quickly Get Ortholog Information." *Getting Genetics Done* [blog], May 11, 2012. http://getting geneticsdone.blogspot.com/2012/05/video-tip-use-ensembl-biomart-to .html.

Ulam, Stanislaw M. "Some Ideas and Prospects in Biomathematics." *Annual Review of Biophysics and Biomathematics* 1 (1972): 277–292.

Ure, Jenny et al. "Aligning Technical and Human Infrastructures in the Semantic Web: A Socio-technical Perspective." Paper presented at the Third International Conference on e-Social Science, 2008. http://citeseerx.ist.psu .edu/viewdoc/summary?doi=10.1.1.97.4584.

US Congress. House of Representatives. "To Establish the National Center for Biotechnology Information," H.R. 5271, 99th Congress, 2nd session, 1986.

US Congress. House of Representatives. "Biotechnology: Unlocking the Mysteries of Disease." Select Committee on Aging, Subcommittee on Health and Long-Term Care (March 6, 1987).

US Congress. Office of Technology Assessment. "Mapping Our Genes: The Genome Projects. How Big? How Fast?" OTA-BA-373 (Washington, DC: US Government Printing Office, April 1988).

Varela, Francisco J. et al. "Autopoiesis: The Organization of Living Systems, Its Characterization and a Model." *Biosystems* 5 (1974): 187–196.

Vilain, Marc. E-mail message to Edward Feigenbaum, October 13, 1987. "Seminar—the Matrix of Biological Knowledge (BBN)." [EAF papers.]

Vokoun, Matthew R. "Operations Capability Improvement of a Molecular Biology Laboratory in a High-Throughput Genome Sequencing Center." Master's thesis, Sloan School of Management and Department of Chemical Engineering, MIT, 2005.

Waddington, C. H., ed. *Towards a Theoretical Biology*. Chicago: Aldine, 1968.

Waldby, Catherine. *The Visible Human Project: Informatic Bodies and Posthuman Medicine*. London: Routledge, 2000.

Wang, Eric T. et al. "Alternative Isoform Regulation in Human Tissue Transcriptomes." *Nature* 456, no. 7721 (2008): 470–476.

Waterman, Michael S. *Introduction to Computational Biology: Maps, Sequences, and Genomes*. London: Chapman & Hall, 1995.

———. "Skiing the Sun: New Mexico Essays." Accessed June 29, 2011. http:// www.cmb.usc.edu/people/msw/newmex.pdf.

Waterman, Michael S. et al. "Some Biological Sequence Metrics." *Advances in Mathematics* 20 (1976): 367–387.

Weinberg, Alvin M. "Criteria for Scientific Choice." *Minerva* 1 (Winter 1963): 159–171.

———. "Impact of Large Scale Science." *Science* 134 (1961): 161–164.

———. *Reflections on Big Science.* New York: Pergamon Press, 1967.

Weinstein, J. "A Postgenomic Visual Icon." *Science* 319 (2008): 1772–1773.

Wells, Terri. "Web 3.0 and SEO." *Search Engine News* (November 29, 2006). http://www.seochat.com/c/a/Search-Engine-News/Web-30-and-SEO/.

Wheeler, D. A. et al. "The Complete Genome of an Individual by Massively Parallel DNA Sequencing." *Nature* 452 (2008): 872–876.

Wiley, Steven. "Hypothesis-Free? No Such Thing." *Scientist* (May 1, 2008).

Wilkes, Maurice V. "Presidential Address: The Second Decade of Computer Development." *Computer Journal* 1 (1958): 98–105.

Wilkinson, Leland. "Dot Plots." *American Statistician* 53, no. 3 (1999): 276–281.

Wilkinson, Leland, and Michael Friendly. "The History of the Cluster Heat Map" (November 18, 2008). Accessed April 25, 2009. http://www.cs.uic .edu/~wilkinson/Publications/heatmap.pdf.

Wise, M. Norton. "Making Visible." *Isis* 97, no. 1 (2006): 75–82.

Wojcicki, Anne. "Deleterious Me: Whole Genome Sequencing, 23andMe, and the Crowd-Sourced Health-Care Revolution." Talk at Science and Democracy Lecture Series, Program on Science, Technology, and Society, Harvard Kennedy School, April 17, 2012. Available at http://vimeo.com/40657814.

Womack, James P. et al. *The Machine That Changed the World: The Story of Lean Production.* New York: Harper Perennial, 1991.

Wong, Yuk-Chor et al. "Coding and Potential Regulatory Sequences of a Cluster of Chorion Genes in *Drosophila melanogaster*." *Chromosoma* 92, no. 2 (1985): 124–135.

World Wide Web Consortium. "OWL Web Ontology Language: Overview" (W3C recommendation, February 10, 2004). http://www.w3.org/TR/owl -features/.

———. "RDF Primer" (W3C recommendation, February 10, 2004). http:// www.w3.org/TR/2004/REC-rdf-primer-20040210/.

Wright, Fred A. et al. "A Draft Annotation and Overview of the Human Genome." *Genome Biology* 2, no. 7 (2001): research0025.1–0025–18.

Yeo, Gene W. et al. "Identification and Analysis of Alternative Splicing Events Conserved in Humans and Mouse." *Proceedings of the National Academy of Sciences USA* 102, no. 8 (February 22, 2005): 2850–2855.

Young, Jake. "Hypothesis-Free Research?" *Pure Pedantry* [blog], May 19, 2008, 11:27 a.m. http://scienceblogs.com/purepedantry/2008/05/ hypothesisfree_research.php.

Yoxen, Edward. "Life as a Productive Force: Capitalizing upon Research in Molecular Biology." In *Science, Technology and Labour Processes*, edited by L. Levidow and R. Young, 66–122. London: Blackrose Press, 1981.

Zand, Tristan Z. "Web 3.0: Back to the Real World/Back to Our Senses." Accessed September 2, 2008. http://www.zzz.ch/bootymachine/web3.0/.

Zaphiropoulos, Peter G. "Circular RNAs from Transcripts of the Rat Cyto-
  chrome P450 2C24 Gene: Correlation with Exon Skipping." *Proceedings of
  the National Academy of Sciences USA* 93, no. 13 (1996): 6536–6541.
Zuckerkandl, Emile, and Linus Pauling. "Molecules as Documents of Evolutionary History." *Journal of Theoretical Biology* 8 (1965): 357–366.

Index